W0254046

ALLE · ZEIT · WACH
1842

Reviews of

74 Physiology, Biochemistry and Pharmacology

formerly
Ergebnisse der Physiologie, biologischen Chemie und experimentellen Pharmakologie

With 60 Figures

Springer-Verlag Berlin
Heidelberg GmbH 1975

ISBN 978-3-662-34770-6 ISBN 978-3-662-35091-1 (eBook)
DOI 10.1007/978-3-662-35091-1

Library of Congress-Catalog-Card Number 74-3674

Originally published by Springer-Verlag Berlin Heidelberg New York in 1975.
Softcover reprint of the hardcover 1st edition 1975

Contents

Indexed in Current Contents

List of Contributors

BLOOM, FLOYD E., Dr., Laboratory of Neuropharmacology, National Institute of Mental Health, Saint Elizabeths Hospital WAW Bldg., Washington, D.C. 20032/USA

DEETJEN, PETER, Prof. Dr., Institut für Physiologie der Universität, Fritz-Pregl-Str. 3, A-6010 Innsbruck/Austria

FOULKES, ERNEST C., Prof. Dr., Depts. of Environmental Health and Physiology, Univ. of Cincinnati, Coll. of Medicine, Cincinnati, Ohio 45267/USA

SILBERNAGL, STEFAN, Univ. Doz., Dr., Institut für Physiologie der Universität, Fritz-Pregl-Str. 3, A-6010 Innsbruck/Austria

Rev. Physiol. Biochem. Pharmacol., Vol. 74

The Role of Cyclic Nucleotides in Central Synaptic Function

F.E. Bloom*

Contents

---

* Laboratory of Neuropharmacology, Division of Special Mental Health Research, IRP, National Institute of Mental Health, Saint Elizabeths Hospital, Washington, D.C. 20032

I. Introduction

A. History and Scope of this Review

Less than two decades ago SUTHERLAND and RALL (1958; RALL et al., 1957) discovered a heat stable factor now known as adenosine 3′,5′-monophosphate, which accumulated in particulate fractions of liver homogenates exposed to epinephrine. This factor activated the breakdown of glycogen by soluble cytoplasmic enzymes. Although epinephrine activates glycogenolysis in intact hepatocytes, the hormone has no such effect on the soluble enzymes. SUTHERLAND and RALL (1960; RALL and SUTHERLAND, 1958, 1961, 1962) proposed that epinephrine might trigger the glycogenolytic response of hepatocytes by activating the synthesis of this nucleotide (cyclic AMP) whose structure (Fig. 1) by then had been shown to contain a unique cyclic phosphate bond (LIPKIN et al., 1959). The polypeptide hormone glucagon also triggers the glycogenolytic response of liver, and activates the synthesis of cyclic AMP. From these and related observations on the actions of other hormones, the concept arose that cyclic AMP functioned as an intracellular "second messenger", synthesized in response to certain hormones and which,

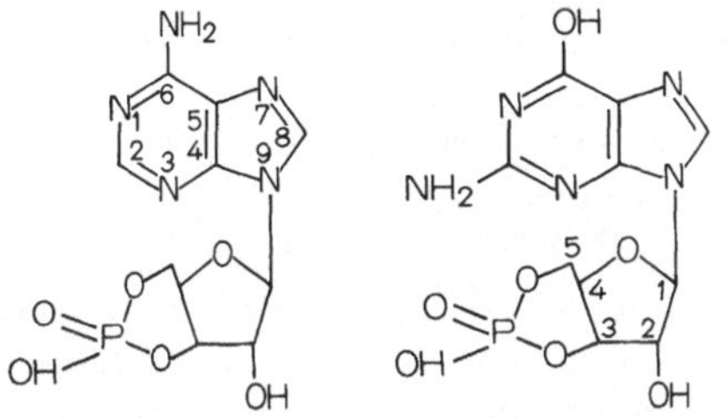

Fig. 1. Schematic, enumerated molecular structures of cyclic nucleotides adenosine 3′,5′-monophosphate (left) and guanosine 3′,5′-monophosphate (right). The purine ring is enumerated in cyclic AMP and the ribose ring enumerated in cyclic GMP; these sites are points at which substitutions have been made in the synthesis of cyclic nucleotide congeners with varying affinity for phosphodiesterases and other cyclic nucleotide dependent enzymes (see text)

by activating the appropriate sequence of enzymes, produced the specific biologic response of the target cell to the hormone (SUTHERLAND et al., 1965).

The subsequent pursuit of the mechanisms which can regulate the levels of cyclic AMP within cells and by which this cyclic nucleotide in turn modulates a multitude of fundamental cellular properties now represents one of the major thoroughfares of biomedical research, involving virtually every species and organ system (see ROBISON et al., 1971).

In the course of this explosive research effort, a second naturally occurring cyclic nucleotide, guanosine-3′,5′-monophosphate (cyclic GMP; Fig. 1), was discovered (ASHMAN et al., 1963). In general, the brain contains relatively high amounts of both these cyclic nucleotides (KLAINER et al., 1962; GOLDBERG et al., 1969; also see RALL and SATTIN, 1970; ROBISON et al., 1971) and the enzymes which synthesize (SUTHERLAND et al., 1962; HARDMAN and SUTHERLAND, 1969) and degrade (CHEUNG, 1970; KAKIUCHI et al., 1971) them. Furthermore, experimental conditions have been described in which the tissue content or synthesis rate of these cyclic nucleotides can be altered upon exposure to substances which are thought to be synaptic transmitters. Among other effective substances, cyclic AMP synthesis in most brain regions seems uniformly responsive to the catecholamines, epinephrine, norepinephrine and dopamine (KLAINER et al., 1962; KAKIUCHI and RALL, 1968b; RALL and SATTIN, 1970; c.f., DALY, 1975) while cyclic GMP levels appear to be more related to acetylcholine (FERRENDELLI et al., 1970; KUO et al., 1972; LEE et al., 1972).

The present review examines the functional significance of cyclic nucleotides in nervous tissue with emphasis upon their possible role in synaptic transmission. In view of the range of effects which can be influenced, particular attention is devoted to the concept that cyclic nucleotides may be second messengers for synaptic transmitters in analogy to their roles in the hormonal response of other tissues. In order to visualize this possibility within the context of the more general phenomena of intercellular chemical communication, a brief overview is first provided of the evolution of the principles of chemical transmission between neurons (Section II.).

The next chapter covers the identifiable macromolecular components of the composite scheme in which cyclic nucleotide levels are regulated and by which their effects are expressed (III.). The subsequent section evaluates the biochemical (IV.A and B), electrophysiological (IV.C) and cytochemical (IV.D) methods by which cyclic nucleotides can be related to specific, identified actions of transmitters such as catecholamines, and other possible transmitters, such as histamine, serotonin and adenosine. To analyze this functional relationship further, the effects of cyclic AMP at catecholamine-mediated synapses in several parts of the central nervous system (V.A, V.B, and V.C) and in the sympathetic ganglia (V.D), are compared to the effects produced by catecholamines in several other tissues (V.D). This coverage provides a comprehensive treatment of the general roles regulated by the catecholamines and the extent to which cyclic AMP may mediate these roles as an intracellular second messenger (V.F). Attention is then drawn to several other lines of evidence which suggest that cyclic nucleotides may influence other aspects of CNS function not ascribable to individual synaptic transmitters (VI). Finally, the conclusions attempt to place these potential cyclic nucleotide-regulated events into a broader perspective of functional neuronal metabolism (VII).

B. Related Sources of Information

The inexorable and rapid expansion of the pertinent literature on cyclic nucleotide regulation and function in nervous tissue challenges every effort to encapsulate facts into a review. Therefore, the present literature search ceased arbitrarily in January, 1975. While the inclusion of every applicable citation was clearly impossible, a broad documentation was attempted. Additional information may be found in the annual series of Advances in Cyclic Nucleotide Research, and in the Journal of Cyclic Nucleotide Research which will commence in early 1975. Furthermore, several fine treatments of the same subject have already appeared (WEISS and KIDMAN, 1969; RALL and GILMAN, 1970; BITENSKY et al., 1973; GREENGARD et al., 1973; MULLEROE and SCHWABE, 1973; BLOOM, 1974; GREENGARD and KEBABIAN, 1974; VON HUNGEN and ROBERTS, 1974; AMER and KREIGHBAUM, 1975; BLOOM et al., 1975; DALY, 1975).

II. The Nature of Chemical Communication between Nerve Cells

For many years, a long, but eventually productive debate centered on the mechanism by which the nerve impulse was transmitted across intercellular junctions. Despite anatomical evidence indicating the separation between contiguous neurons and their post-junctional cells, and despite the demonstration by LOEWI (1921) that the cardiac vagal impulse was transmitted chemically, the concept that an unbroken physical wave of electrical excitation accounted for the rapid transmission from nerve to skeletal muscle retained firm supporters well into the 1950's (see ECCLES, 1964).

A. Transmission between Nerve and Muscle

With the introduction of the microelectrode for examination of transmembrane potentials (LING and GERARD, 1949), the arrival of electrical activity in the nerve terminal could be separated from the onset of the wave of depolarization in the muscle fiber (see KATZ, 1966). This synaptic delay period clearly indicated that the electrochemical process by which excitation propagates within the axon did not spread directly to the muscle. In fact, the junctional zone (or end-plate) of muscle appeared not to be electrically excitable (KUFFLER, 1949). Furthermore, the muscle response could be blocked by curare, lending credence to the hypothesis that nerves transmit information to muscle by secretion of acetylcholine (ACh) (DALE et al., 1936). Subsequent efforts (see KATZ, 1966) revealed that ACh was released presynaptically, that applied ACh produced an effect on the postsynaptic muscle identical to the nerve impulse, and that the nerve released sufficient ACh to account for the changes in muscle membrane properties observed during junctional transmission. In addition, the sub-threshold end-plate potentials could be potentiated in amplitude and duration by inhibitors of the acetylcholinesterase (DEL CASTILLO and KATZ, 1957) known to be present at the junction (COUTEAUX, 1955) by histochemical localization (KOELLE and FRIEDENWALD, 1949).

The weight of these collective electrophysiological, biochemical, pharmacological and histochemical experiments established conclusively the chemical nature of nerve-muscle transmission, and the identity of ACh as the excitatory transmitter. These experiments also indicated that secretion of the transmitter was an essential function of the nerve (as predicted by SHERRINGTON, 1925). Both the secretory event and the response triggered by the transmitter at its postsynaptic receptor are ion dependent processes: Ca^{++} is required for ACh secretion by the excited nerve terminal and Ca^{++} is also required for the spread of excitation from the sub-synaptic region into the contractile elements of muscle (V.D4; see reviews by GERGELY, 1974; HUXLEY, 1974; WEBER and MURRAY, 1973). Furthermore, the initial excitatory potential generated within the subjunctional zone by combination of the transmitter, ACh, with its receptor, is analogous to the ionic process which propagates activity along the axonal membrane: a large but sequential increase in conductance of the membrane to sodium and then potassium results in a depolarization. The extent of the depolarization in the muscle depends on the amount of ACh secreted and on the composition of the extracellular fluid as would be anticipated by the general ionic membrane hypothesis (FATT and KATZ, 1951).

B. Transmission between Nerve Cells

Based on the interpretation of the chemical and ionic basis of junctional transmission in the peripheral nervous system (see GINSBORG, 1967; WEIGHT, 1971), and on similar phenomena observed in the analysis of central synaptic transmission (see ECCLES, 1964), the generalization arose that all chemical neurotransmitter substances produce their effects on postsynaptic neurons by changing ionic permeability. Under this generalization, the basic mechanism accounting for the response to the synaptic transmitter is an increase in permeability of the subsynaptic zone to one or more ions which shifts the potential across the membrane toward the equilibrium potential for the affected ions (at equilibrium an increase in permeability results in no net movement because the electrochemical gradients driving the ions are balanced). Thus, excitatory transmitters would activate ionic conductances leading to depolarization, while in the opposing process of inhibition, transmitters should activate ionic conductances which lead to hyperpolarization. Inhibition could occur without change in membrane potential, simply as a result of increased ionic conductances sufficient to shunt excitatory receptor mechanisms. On the basis of extensive investigations of inhibitory mechanisms in the cat spinal cord, most inhibitory synaptic potentials would seem to involve increased Cl^- permeability (see ECCLES, 1964; WERMAN, 1972; CURTIS and JOHNSTON, 1974). In some cases, the same transmitter substance can produce opposite electrophysiological effects through alteration of the same ionic conductance (see GARDNER and KANDEL, 1972). This may occur because the equilibrium potential for the responsible ion in the responding cell varies markedly due to differences in internal ion concentration and in ionic permeability. Thus, gamma amino butyrate (GABA) can hyperpolarize motoneurons in the frog spinal cord, while depolarizing the primary afferent terminals of dorsal root ganglion cells (BARKER et al., 1971; DAVIDOFF, 1972).

Recent evidence has suggested that some membrane potential changes observed in sympathetic ganglia (see WEIGHT, 1974), retina (BAYLOR and FUORTES, 1970) and in the central nervous system (SIGGINS et al., 1971 a, b; ENGBERG and MARSHALL, 1971; KRNJEVIC et al., 1971; HOFFER et al., 1973; OLIVER and SEGAL, 1974; ENGBERG et al., 1974) may result from mechanisms in which the transmitter substance decreases resting ionic permeability rather than increasing it as in the previous examples (see also Section V.). In fact, the synaptic mechanisms with which cyclic nucleotide actions are mostly closely associated arise from studies of these same regions.

C. Identification of Synaptic Transmitter Substances

Identification of the chemical substance which a particular set of synaptic junctions secretes is an essential step in the determination of the molecular mechanism of information transmission. From the studies on the chemical transmission process in peripheral innervated tissues, rigorous criteria were developed for the identification of neurotransmitter substances. Although various authors have individual versions (see PATON, 1958; WERMAN, 1972; BLOOM, 1968, 1974; CURTIS and JOHNSTON, 1974), the criteria usually require that: (1) the substance can be shown to be present in the terminals of the nerves under study; (2) that it is released from these nerves when they are stimulated; and (3) that the pure substance, when applied experimentally, produces an effect identical to nerve stimulation. These criteria are rather easily applied to peripherally innervated tissues in which the nerves could be isolated by dissection for purposes of chemical assay and electrical stimulation, and where the end organ could be perfused during nerve stimulation for detection of the chemicals released. Within the central nervous system, however, the anatomical complexity of individual synaptic projections and inadequate sensitivity for chemical detection of transmitter substances prevents the clearcut satisfaction of these criteria. As a result, the presence of a chemical within the terminals of a specified pathway can be firmly established only by cytochemical tests for transmitter substances or their synthetic enzymes (see BLOOM, 1973; HÖKFELT, 1973) or by the loss of a specific transmitter substance after a single neuronal population is destroyed (see WERMAN, 1973).

Due in part to the uncertainties in identifying the transmitter contained within a given central synaptic projection and in the detection of its release, additional secondary criteria were evolved to improve the accuracy of comparing the actions of the natural transmitter with those of candidate substances. To satisfy these secondary criteria (see WERMAN, 1966, 1973), the candidate transmitter substance must act on the membrane potential in the identical manner as the natural pathway, namely by producing the identical change in conductance for the same ions. If identical effects are produced, both substances should then drive the membrane toward the same ionic equilibrium potential. A less rigorous, but more often applied test of identity of action depends upon pharmacological tests: selective blockade or potentiation of the effects of the candidate substance (e.g., specific receptor antagonists or inhibitors of catabolism) should produce identical blockade or potentiation of the effects of pathway activation. While none of the criteria applied individually are considered adequate for identification of the transmitter substance, the concordance of multiple tests provides a strong case for the identification.

D. The Interaction between Neurotransmitters, Hormones and their Receptors

A logical first step in assessing the roles which cyclic nucleotides may play in nervous tissue is to consider other tissues (see ROBISON et al., 1971), in which cyclic AMP and cyclic GMP function as mediators of the action of certain circulating hormones. In one of the earliest theoretical discussions of chemical communication between cells, HUXLEY (1935) proposed that hormones were molecules which transfer information between sets of cells. In this general context, neurotransmitters can easily be conceived as a subset of hormones characterized by a limited area of receptivity and a short period of activity after secretion. These properties contrast with the diffuse area of receptivity and relatively long time of action which characterizes circulating hormones such as insulin. In fact, some substances, such as the catecholamines, epinephrine and norepinephrine, can act in both ways: as circulating hormones when they are secreted from the adrenal medulla, and as neurotransmitters when they are secreted from nerves at specific points within the peripheral or central nervous system.

Delimited fields of reactivity and brief periods of reaction are essential to rapid and discrete informational processing within the central nervous system. In fact, neurotransmitter substances which do not act rapidly in discrete point-to-point circuits may therefore seem less like "transmitters" than hormones. Nevertheless, the analysis of the mechanisms by which hormones and neurotransmitters exert their actions reveals many common factors, the primary common denominator being the concept of the receptor. In both cases, cells which respond to a hormone are assumed to possess a "receptor" which can recognize the transmitter, bind it specifically, and thereby initiate the sequence of actions which leads to the response. The receptor for a neurotransmitter is characterized partially by the molecules which can selectively simulate or antagonize transmitter actions and partially by the change in conductance to ions which leads to the biophysical changes in membrane properties, generally considered more simply as excitation or inhibition.

In the case of circulating hormones which lead to biochemically measured events (such as glycolysis, lipolysis, or hormone secretion), the abstract hormone receptor concept may soon be displaced by a description of the molecular events by which the hormone triggers the response of the target cells (see CUATRECASAS, 1974, 1975). Thus, for many hormones, the earliest measurable change induced by specific hormone binding now appears to be control of the synthesis of either cyclic AMP or cyclic GMP (ROBISON et al., 1971; CUATRECASAS, 1974, 1975).

However, in the case of neurotransmitters, the molecular mechanism of the response in which the binding of transmitter by receptor specifically alters the ionic conductivity of the membrane, is still abstract. If nerve cells communicate mainly by secretion of chemical transmitters and if neurotransmitters can be regarded as a specialized group of hormones acting rapidly within the locale of their secretion, is it reasonable to assume that some of the mechanisms by which hormones trigger events in peripheral target cells may also apply to the action of certain transmitter substances in the nervous system? To formulate a comprehensive answer to this fundamental question, we next examine in greater detail the components of a hormone-triggered event and the way in which the mediation of such analogous events within the nervous system by cyclic nucleotides might be established.

III. The Components of a Cyclic Nucleotide-Mediated Hormone-Triggered Event

In many tissues, a hormone triggers receptive cells by a multi-step process through the intermediation of cyclic nucleotides. This process requires several macromolecular components which are illustrated schematically in Fig. 2: a) the *receptor* for the hormone or transmitter substance which directly alters the activity of b) the *nucleotide cyclase* (adenylate cyclase or guanylate cyclase) to raise the intracellular concentration of the cyclic nucleotide. The intracellular cyclic nucleo-

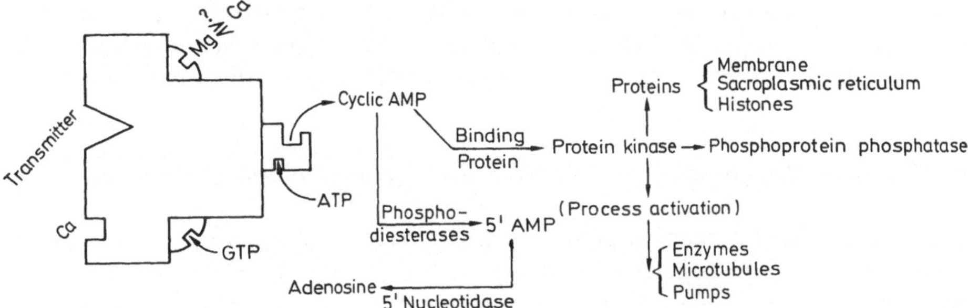

Fig. 2. Schematic portrayal of the macromolecular components of the multi-enzyme sequence by which hormones and transmitters can evoke their selective responses from target cells. The sequence begins at left with hormone recognizing receptors linked to adenylate cyclase. This receptor-cyclase complex contains modulatory sites sensitive to Ca^{++}, Mg^{++}, guanyl nucleotides, and F^-, as well as a substrate binding site for catalysis of ATP. The cyclic AMP synthesized then is either catabolized by phosphodiesterase or thought to influence other sensitive systems within the cell by regulating the activity of specific protein kinases. These kinases may transfer phosphates onto structural or enzymatic proteins which regulates their activity; such a process is then terminated by a phosphoprotein phosphatase. Activation by cyclic AMP-sensitive protein kinases may also activate intracellular processes such as secretion, microtubular assembly, or ionic pumps through substrate proteins which so far remain uncharacterized. This scheme is based upon multiple sources (see DALY, 1975; GREENGARD and KEBABIAN, 1974; RODBELL, 1975; and text)

tide concentration influences the activity of a general class of phosphotransferases, termed c) *protein kinases* (WALSH et al., 1968; KUO and GREENGARD, 1969a, b; LANGAN, 1972). These enzymes can phosphorylate intracellular proteins such as enzymes or structural proteins, which presumably alters their physiological properties. One of the enzymes which may be activated in this way is d) a *phosphoprotein phosphatase* (but see MAENO et al., 1975 and VI.B., below) which could then restore the substrate molecule to its original state by the removal of the same phosphate group (UEDA et al., 1973). Finally, the cyclic nucleotides may be degraded by e) specific *3',5'-phosphodiesterases* (see AMER and KREIGHBAUM, 1975).

These five major components comprise a chained sequence of events which could explain a substantial portion of the mechanism by which hormones, and possibly neurotransmitters, evoke their effects. When viewed in the perspective of communication efficiency, the scheme offers the opportunity for signal amplification since the hormone can catalyze the formation of the cyclic nucleotide, and catalytic amounts of the cyclic nucleotide can activate the protein kinases. At the same time the multi-step complex of reactions also offers numerous opportunities for interaction with the intracellular systems supplying needed substrates, and regulatory co factors, such as Ca^{++} and Mg^{++}. These latter interactions are therefore also capable of modifying (or interfering with) various steps in the hormone response.

The nucleotide-mediated system of hormone responses may seem initially to be complicated and potentially subject to the hazards of fluctuating cellular metabolism. However, the rather extensive documentation of this system in liver and fat cell metabolism (see LANGAN, 1970, 1972) suggests that any reticence in our acceptance of this scheme in nervous tissue should be based upon the caution appropriate to incompletely examined phenomena rather than skepticism.

A. Hormone Receptors: Physiological and Pharmacological Properties

The first step in the chain of events by which the hormone or neurotransmitter begins to act is the process of binding to the specific receptors of the target cell. For many years the only method by which this receptor could be inferred was to measure the potency of the hormone, or of structurally modified agonists or antagonists of the hormone, in the elicitation of the response event of the target cell. In recent years, however, methods have evolved for the partial purification of the receptors from membrane fragments of some receptive cells (AURBACH et al., 1974; LEFKOWITZ et al., 1974). The status of this art and a critical enumeration of the substantial pitfalls involved in their interpretation has recently been reviewed by CUATRECASAS (1974, 1975).

Because of the intimate relationship between many of the hormonally evoked responses and the activation of adenylate cyclase, the assay of the enzyme activity associated with hormonal stimulation has been one of the more helpful tools in attempting to purify the receptors for glucagon (RODBELL, 1975), and insulin (CUATRECASAS, 1975). Although a catecholamine-sensitive preparation of brain homogenates was described early in the course of the investigations into the role of cyclic AMP in the central nervous system (KLAINER et al., 1962), most efforts to extend those studies were disappointing because the degree of activation of brain adenylate cyclase by catecholamines was relatively small (less than 1 order of magnitude), compared to liver or adipocytes (ROBISON et al., 1971).

In the hepatocyte and adipocyte systems in which hormonal responsiveness is retained under broken cell conditions, two interpretations have arisen to explain activation of cyclase activity by hormones and to account for the fact that more than one hormone is able to activate the adenylate cyclase of a given cell type. In one view (see RODBELL, 1975), the adenylate cyclase possesses a component which faces the extracellular space and which can "discriminate" the presence of one hormone from another, reacting only with those hormones which exhibit the specific three dimensional structure required (RODBELL et al., 1971). When such a hormone occupies the appropriate part of a properly primed receptor complex, the conformational changes which occur result in enzyme activation. This view has, however, continued to evolve under vigorous probing by the Rodbell group and others (LONDOS et al., 1974; RODBELL, 1975; also see LEFKOWITZ, 1974; SPIEGEL and AURBACH, 1974; BIRNBAUMER and YANG, 1974a, b; BIRNBAUMER et al., 1974) as the nature of the co-factors required for proper priming of the receptive enzyme complex have become more refined. These latter developments appear to have made possible the precise type of kinetic analysis needed to relate receptor binding to activation of the adenylate cyclase (RODBELL, 1975). Nevertheless, the general connotation of this model would be that the hormone receptor is an integral part of the enzyme in some form of complex such that when the receptive portion binds the hormone, the catalytic site is subjected to conformational changes or is, perhaps, "de-regulated" by the release of a currently undetectable regulatory component.

Recently, CUATRECASAS (1974, 1975) has proposed an alternative concept (see also PERKINS, 1973) in which the hormonally receptive component and the catalytic enzyme component are considered as independent protein units, migrating independently within the fluid mosaic of the plasma membrane; when the hormone occupies the receptor, the conformation of the receptor changes and this results in its rapid coupling to the catalytic component of the adenylate cyclase, and activation. Under this concept it might be possible for more than one hormone or transmitter to bind to their specific class of membrane receptors to activate the cyclase. Moreover, each agonist could thereby form a complex with its receptor which could, in turn, activate several membrane-related processes including adenylate cyclase. If such a purified receptor binding component could be isolated and shown to activate adenylate cyclase, it would permit an evaluation of one of the fundamental unknowns of the cyclic nucleotide mediated systems: do the hormones which activate these systems do so only through the formation and subsequent various actions of the cyclic nucleotides or do the hormones produce many subtle independent effects, only one of which is the overt formation and activation of the second messenger scheme?

B. Adenylate Cyclase

Among the initial set of observations by SUTHERLAND and RALL was the description of the reaction catalyzed by the enzyme they termed "adenyl cyclase" (SUTHERLAND et al., 1962; RALL and SUTHERLAND, 1962; KLAINER et al., 1962). In this reaction, ATP in the presence of Mg^{++} is converted to cyclic AMP and pyrophosphate; in broken cell preparations, the enzyme is associated mainly with the particulate fractions, and the basal activity can be stimulated by addition of F^- and by a variety of hormones (see below). Under appropriate reaction conditions, Greengard and his colleagues (see HAYAISHI et al., 1971) demonstrated the reaction to be reversible, with the equilibrium constant slightly in favor of ATP production.

In nervous tissues, the regional concentration of the enzyme varies substantially; the overall activity is higher in cerebellum and cerebral cortex (WEISS and COSTA, 1968; also see KLAINER et al., 1962). On subcellular fractionation of the brain, the activity was reported to be associated with fractions enriched in nerve terminal membranes, especially when they are exposed to hypotonic media (DE ROBERTIS et al., 1967). However, it seems quite apparent that the activity present in broken cell membranes is only a relatively small component of the total present in intact tissues. When partially purified, both neurons and glia exhibit enzyme activity (PALMER, 1973). While transmitter responsiveness is generally lost or severely reduced by fractionation, activation by NaF is retained (SUTHERLAND et al., 1962). Furthermore, brain generally exhibits a relatively high "basal" level of enzyme activity, but such activity cannot necessarily be considered to be transmitter-independent (see PERKINS, 1973) and could result in part from manipulations during fractionation.

Although the brain adenylate cyclase has not been purified to the extent of other membrane sources, it seems clear that enzyme from brain also exhibits an absolute requirement for Mg^{++} which may be partially related to the attachment of the ATP substrate in the proper fashion (PERKINS and MOORE, 1971). In some cases the Mg^{++} requirement can be met by either Mn or Co, and generally not by Ca, Zn, or Cu (see PERKINS, 1973), but until the enzyme is purified from a variety of sources, it is unclear just what these metallic ion interactions imply regarding the function or location of the enzyme.

Recently, the use of guanylylimidodiphosphate (LEFKOWITZ, 1974; LONDOS et al., 1974; SPIEGEL and AURBACH, 1974; RODBELL, 1975) has partially clarified the presence of a guanine-riboside sensitive component to the catalytic portion of the liver and erythrocyte adenylate cyclases, a site presumed normally to be occupied by GTP. Thus, in addition to sites capable of binding the ATP substrate, the adenylate cyclase must also contain at least one site of required interaction for a divalent metal as well as a site for GTP in order for the hormone to be able to stimulate activity (see RODBELL, 1975). The means by which NaF is also able to activate activity is unclear, but it does so in almost all cases even when hormonal sensitivity has been lost. Other agents, especially cholera toxin (CUATRECASAS, 1973; FIELD, 1974) can activate both basal and catecholamine-stimulated adenylate cyclase, possibly by facilitating the association between catalytic components with hormone reactive components (CUATRECASAS, 1974).

Recent studies by Birnbaumer and his colleagues (BIRNBAUMER and YANG, 1974a, b; BIRNBAUMER and YANG, 1974a, b; BIRNBAUMER et al., 1974a, b) suggest that the partially purified adenylate cyclases of several tissues, (but not brain) are regulated by the diphosphates and triphosphates of adenosine, guanosine and possibly inosine, in such a way that the diphosphates and triphosphates have opposing (i.e., activating or inhibiting) actions, Furthermore, BILZEKIAN and AURBACH have reported (1974) that GTP can augment the response of turkey erythrocytes to activation of adenylate cyclase by isoproterenol and that this effect of GTP was specific for the activation by the catecholamine. If these results in non-neuronal tissues can ultimately be shown to apply to nervous tissue, the expression of hormonal responsiveness by a given cell type becomes staggeringly complex, and intimately dependent upon the entire metabolism of the cell.

C. Guanylate Cyclase

Although naturally occurring cyclic GMP was discovered shortly after the discovery of cyclic AMP, progress in the establishment of an analogous functional scheme for cyclic GMP in the mediation of hormonal actions has been substantially slower. Two major reasons account for this "cyclic nucleotide gap": (1) the difficulties, until recently, of demonstrating that hormones could alter tissue levels of cyclic GMP and (2) the lack of a sensitive assay system capable of detecting the nucleotide (see GOLDBERG et al., 1973). However, within the past 5 years, several groups have rapidly advanced the state of research on cyclic

GMP by determining its relatively high concentration in lung, brain, intestine and thyroid (GOLDBERG et al., 1969; MURAD et al., 1971; KUO et al., 1972; FERRENDELLI et al., 1970, 1972, 1973; KIMURA et al., 1974). Furthermore, the presence of relatively high amounts of guanylate cyclase in brain (HARDMAN and SUTHERLAND, 1969) and other organs (WHITE and AURBACH, 1969) and the discovery of some cyclic GMP sensitive protein kinases in lobster muscle (KUO and GREENGARD, 1970b), and in mammalian smooth muscle (CASANELLIE et al., 1974) and brain (HOFMANN and SOLD, 1972; KUO, 1974) strengthens the hypothesis that cyclic GMP could also serve as an intracellular second messenger analogous to cyclic AMP (see GOLDBERG et al., 1970, 1973a, b; GOLDBERG, 1975). In fact, based upon the relative changes in levels of cyclic GMP and cyclic AMP in tissues which are subject to transmitter regulation by more than one hormone, GOLDBERG has suggested that the two nucleotide systems may interact as dual opposing intracellular regulators akin to the oriental concept of Yin and Yang. In the perfused heart (GEORGE et al., 1970), acetylcholine can elevate cyclic GMP levels through an atropine-sensitive muscarinic receptor (GOLDBERG et al., 1973a); this response is the prototype of hormonal responses to neurotransmitters, exhibited by guanylate cyclase in innervated tissues, including brain (see FERRENDELLI et al., 1970, 1972, 1974; MAO et al., 1974a, b, c).

Unlike adenylate cyclase, guanylate cyclase activity in tissue homogenates tends to become enriched in the non-membranous or soluble fractions (HARDMAN and SUTHERLAND, 1969; WHITE and AURBACH, 1969), although GOLDBERG (GOLDBERG et al., 1973a) has interpreted the available data to indicate that some of this distribution may indicate translocation of the enzyme from membranes in which it is less firmly attached than adenylate cyclase. The precise location of the enzyme becomes crucial to the question of how and when the nucleotide levels are altered during a specific hormonal response.

A second characteristic property of the guanylate cyclase is the order of magnitude of increase in activity obtained when Mn^{++} is substituted for Mg^{++} (HARDMAN and SUTHERLAND, 1969), and the additional observation (SCHULTZ et al., 1973) that Ca^{++} in the presence of low Mn^{++} (less than 0.5 mM) can produce an additional several fold increase in guanylate cyclase activity (NAKAZAWA and SANO, 1974). These interactions are essentially unchanged after a 30-fold purification of brain guanylate cyclase (NAKAZAWA and SANO, 1974); with this partially purified preparation, Mn^{++} seems to be absolutely required for basal activity.

A third and possibly more fundamental property contrasting the guanylate cyclase from adenylate cyclase has been the lack of any agent (such as transmitters or F^-) which can activate enzymic activity (SCHULTZ et al., 1969) in cell free systems (but see RUDLAND et al., 1974). As a result, pharmacological studies on cyclic GMP level changes in nervous tissues have required that substantial doses of agonist be administered to the living animal or to a whole perfused organ (see GOLDBERG et al., 1974; GOLDBERG, 1975; FERRENDELLI et al., 1970, 1972, 1973; MAO et al., 1974a, b, c). However, LEE et al. (1972) established that slices of rabbit cerebral cortex exhibit 2–3 fold increases in cyclic GMP content upon exposure to 1 μM bethanechol or acetylcholine and that this effect is antagonized by atropine.

D. Chemical Reactions of Cyclic Nucleotides

1. Activation of Phosphotransferase Reactions

It is perhaps only moderately ironic that the last enzyme in the sequence by which epinephrine and glucagon are thought to act, namely protein kinase, was first observed even before cyclic AMP was identified (see LANGAN, 1972), and that studies of the phosphorylation of the enzyme phosphorylase provided the initial discoveries leading to the recognition of cyclic AMP formation. Based upon his studies of the high energy content of the cyclic phosphate bond in cyclic AMP (HAYAISHI et al., 1969; RUDOLPH et al., 1971) GREENGARD (KUO and GREENGARD, 1969a, b) made a systematic examination of the possible reactions in which cyclic AMP might participate as an adenylating agent and as part of this study described the presence in brain and several other tissues of a cyclic AMP-sensitive enzyme which would transfer phosphate from ATP to a variety of synthetic substrate proteins, such as histones and casein.

In brain, cyclic AMP-dependent protein kinases (see MIYAMOTO et al., 1969a, b, 1971; WELLER and RODNIGHT, 1970) are associated with the nerve ending membrane fractions of homogenates (MAENO et al., 1971); these membranes can also serve as substrates for the phosphate transfer reaction (JOHNSON et al., 1971, 1972). In addition, the nerve ending-enriched fractions of brain also contain a phosphoprotein phosphatase (MAENO and GREENGARD, 1972). Recently, GREENGARD and his colleagues have described in detail the properties of cyclic AMP-activated phosphate transfer reactions to two species of synaptic membrane protein (Protein I and Protein II) identifiable on acrylamide gel electrophoresis (UEDA et al., 1973). The substrate proteins, whose functions have not yet been determined, are rapidly phosphorylated in the presence of micromolar concentrations of cyclic AMP. One of the substrates, (Protein II), is also rapidly dephosphophorylated; a protein with similar properties on acrylamide gel electrophoresis could also be seen in homogenates of other organs. However the synaptosomal membrane protein substrate, Protein I, seems thus far unique to brain. Of particular interest to the question of synaptic functions of cyclic AMP related systems is the finding by UEDA et al. (1973) that divalent cations can severely influence the direction and extent of the enzyme reaction. Thus, while basal and cyclic AMP-stimulated activity of phosphate transfer to either Protein I or II occurs with 2–4 mM Mg^{++}, or with similar concentrations of Co^{++} or Mn^{++}, substitution of Ca^{++}, Fe^{++}, Cu^{++} or Ba^{++} for the Mg^{++}, resulted in loss of cyclic AMP activation of the protein kinase. When Zn^{++} replaced Mg^{++}, or with 10 mM Ca^{++}, the phosphorylation of Protein II was now, in fact, inhibited by cyclic AMP rather than stimulated.

These ionic effects may indicate some of the potential macromolecular changes involved in the activation of the protein kinases by cyclic AMP. The present interpretation of available data suggests that the protein kinase activation involves the binding of cyclic AMP to an inhibitory regulator normally attached to the catalytic portion of the enzyme. In the presence of cyclic AMP the regulator is thought to dissociate from the catalytic component, thereby permitting activity to proceed (see LANGAN, 1972). The regulatory protein binds cyclic AMP firmly, and was originally studied because of its property as a cyclic AMP-binding protein. Later, the binding protein was observed to inhibit cyclic AMP-independent protein kinases from cell-free homogenates; when cyclic AMP was re-added to "inhibited protein kinase" enzyme activation resulted (see LANGAN, 1972). In addition to the cyclic AMP-binding regulatory protein, an additional natural protein inhibitory factor which can react with the catalytic portion of protein kinase has been discovered in high concentration in brain. This inhibitor does not react with cyclic AMP, but instead binds directly to the catalytic subunit, inactivating it, as does the normal regulatory (or cyclic AMP-binding) protein (ASHBY and WALSH, 1972). Thus, the natural inhibitor could be a degraded regulatory protein which retains affinity for the catalytic unit but not for cyclic AMP.

In addition to activating the transfer of phosphates to proteins structurally related to synaptic membrane fractions, cyclic AMP has also been reported to activate the transfer of phosphate to tubulin (GOODMAN et al., 1970; EIPPER, 1974), to the basic protein of myelin (CARNEGIE et al., 1973; MIYAMOTO and KAKIUCHI, 1974) to cerebral ribosome fractions (SCHMIDT and SOKOLOFF, 1973) and to other large proteins isolated from aplysia ganglioneurons (LEVITAN and BARONDES, 1974; LEVITAN et al., 1974).

One indication of the rapid progress being made with phosphotransferase reactions, their possible functional significance and their evolving conceptualization derives from the work on tubulin. Although bovine brain tubulin can act as a substrate for a protein kinase activity present in a crude homogenate preparation (GOODMAN et al., 1970), the relative activity is low. Tubulin purified from porcine brain (see SOIFER, 1975) and chick brain (SLOBODA et al., 1975) also possesses endogenous protein kinase activity. Although the porcine enzyme can be converted to a form which is fully active without cyclic AMP (SOIFER, 1975) the chick enzyme is extremely sensitive to cyclic AMP when assayed with an endogenous protein substrate, termed by SLOBODA et al. (1975) as microtubule-associated protein. This protein, which is clearly distinct from tubulin itself, constitutes less than 3% of crude microtubular protein, yet it accepts more than 65% of the phosphate transferred in the presence of cyclic AMP (half-maximal activation at 2×10^{-7} M cyclic AMP; SLOBODA et al., 1975). SLOBODA et al. (1975) suggest that this substrate protein may be an ATPase with motile properties which is positioned along assembled microtubules.

Although there have been some difficulties in establishing a specific cyclic GMP-activated transfer of phosphate to mammalian brain protein (KUO and GREENGARD, 1973; KUO, 1974), CASNELLIE and GREENGARD (1974) have recently demonstrated that cyclic GMP activates phosphotransferase activity in smooth muscle preparations in which acetylcholine elevates cyclic GMP levels. The definitive demonstration that the alteration of protein phosphorylation state can change the functional properties of the synaptic substrate proteins is still lacking. However, striking correlation of protein dephosphorylation and membrane property changes has been observed in toad bladder (DE LORENZO et al., 1973; DE LORENZO and GREENGARD, 1973) and a cyclic AMP-dependent phosphoprotein phosphatase can rapidly dephosphorylate synaptic membrane Protein II (MAENO et al., 1975). From these and other data (see below), cyclic nucleotides within cells can be inferred to interact with a substantial variety of protein substrates which could clearly influence cell properties in a rapid and reversible fashion.

2. Catalysis by Phosphodiesterases

All cells which contain the enzymatic capacity to synthesize cyclic nucleotides also possess the only known class of enzyme able to catabolize these nucleotides, namely 3′,5′-phosphodiesterases (see APPLEMAN et al., 1973). These enzymes split the 3′ leg of the cyclic phosphate bond yielding the 5′ purinoside monophosphate. Although the initial studies on this enzymatic activity (SUTHERLAND and RALL, 1962) did not characterize the components of the reaction, recent investigations (see BEAVO et al., 1970) provide extensive documentation that there are multiple molecular forms of the enzyme (MONN and CHRISTIANSEN, 1971) with rather crucial individual characteristics important to the question of synaptic function. Fortunately there have been in depth reviews of this general class of enzyme recently (CHEUNG, 1970; MONN and CHRISTIANSEN, 1971; APPLEMAN et al., 1973; AMER and KREIGHBAUM, 1975; KAKIUCHI, 1973, 1975) so that this section can deal mainly with topics of greater relevance to functional studies of the nervous system.

From the earliest studies, it was clear that the brain contained a relatively high concentration of total phosphodiesterase activity (SUTHERLAND and RALL, 1962), much of which was associated with the same subcellular fractions rich in adenylate cyclase (DE ROBERTIS et al., 1967). Both enzymes are actively transported in chick axons (BRAY et al., 1971). The relative distribution of the total enzyme activity differs markedly from region to region in the brain (WEISS and COSTA, 1968; BRECKENRIDGE and JOHNSTON, 1969) in a manner such that the ratio of synthetic to degradative activities bore a close relationship to the relative tissue levels of cyclic AMP (WEISS and COSTA, 1968; see also WEISS and KIDMAN, 1969). During brain development, phosphodiesterase is associated with nerve terminals (GABALLAH and POPOFF, 1971).

However, as these studies progressed, many reports indicated that the total phosphodiesterase activity of brain was in fact made up of a variety of enzymes (THOMPSON and APPLEMAN, 1971; MONN and CHRISTIANSEN, 1972) which varied with regard to substrate specificity and affinity, molecular weight, migration on gel chromatography or acrylamide gel electrophoresis, heat stability, and sensitivity to cations (see AMER and KREIGHBAUM, 1975; also APPLEMAN et al., 1973; APPLEMAN and TERASAKI, 1975; KAKIUCHI et al., 1972; STRADA et al., 1974; KAKIUCHI, 1975; UZUNOV and WEISS, 1971, 1972; UZUNOV et al., 1974; WEISS, 1972). These substantial variations in physico-chemical properties of the enzyme take on functional importance for two reasons: (1) The use of phosphodiesterase inhibitors to decrease catalysis has been one of the major tactics employed to obtain supportive evidence that the effects of a hormone are produced through a cyclic nucleotide system. However, if there are multiple species of phosphodiesterase with each having specific cellular locations, kinetics, substrates and drug sensitivities, then drug interactions must be defined in these terms to obtain functional interpretations. (2) Most of the assay systems which were used to characterize the location of enzyme activity within regions and subcellular fractions of the brain did not use methods which would have discriminated between the various species of phosphodiesterase isozymes.

The presently available data seem to indicate that at least two major forms of phosphodiesterase exist in brain, although the lack of a uniform nomenclature for correlating the different methods of enzyme separation with the relative importance of substrate concentrations and co-factors, makes comparison of results from the various laboratories difficult. Based upon substrate affinity and metallic cation activation, one phosphodiesterase shows relatively high affinity for cyclic GMP and is activated by a separate Ca^{++}-sensitive modulating protein (KAKIUCHI et al., 1971, 1973; LIN et al., 1974), while a second phosphodiesterase has a low Km for cyclic nucleotides, is more specific for cyclic AMP, shows relatively no sensitivity to added Ca^{++}, and may even be directly influenced by cyclic AMP levels (APPLEMAN et al., 1973; APPLEMAN, 1975). In chick embryonic fibroblasts, the cyclic AMP and cyclic GMP phosphodiesterases appear to be under separate genetic control (RUSSEL and PASTAN, 1974).

However, there are more than 2 protein peaks with phosphodiesterase activity which can be resolved from the soluble supernatant fractions of brain (UZUNOV and WEISS, 1972; STRADA et al., 1974) or cultured glioma cell lines (UZUNOV et al., 1974) and each peak shows preferential sensitivity to pharmacological inhibitors. Furthermore, despite the apparent relative specificity of each of the two major enzyme fractions, both isoenzymes will catabolize either cyclic nucleotide, while each cyclic nucleotide can influence the rate at which the enzymes catabolize the other as substrate (see AMER and KREIGHBAUM, 1975). In addition to these complex interrelationships between phosphodiesterase isozymes *in vitro*, a variety of natural metabolic intermediates (CHEUNG, 1970) and short chain fatty acids (SCHROEDER, 1974) can regulate enzyme activity also, including such effects as the chelation of required amounts of Ca^{++} by added ATP (TESHIMA et al., 1974).

Finally, it should be recognized that an incredible spectrum of pharmacological compounds inhibits phosphodiesterase activity (see WEINRUB et al., 1972) including a variety of compounds known to affect behavior either as stimulants or tranquilizers (BEER et al., 1972; SCHULTZ, 1974). However, the observation that a drug with behavioral actions can also influence the activity of this relatively crucial catabolic enzyme need not be regarded as conclusive proof that cyclic nucleotides are involved in the behavioral effect (see HONDA and IMAMURA, 1968; NAHORSKI et al., 1973; FUXE and UNGERSTEDT, 1974; KLAWANS et al., 1974; ARBUTHNOTT et al., 1974). Therefore, the close interrelationships between the phosphodiesterase enzymes with regard to mutual influential effects of the natural substrates, and their close dependence upon substrate concentrations and Ca^{++} levels cannot be too strongly emphasized in attempts to interpret drug actions objectively. While it is possible that some of these complicated interactions may not occur in intact cells in vivo, APPLEMAN (APPLEMAN et al., 1973; RUSSELL et al., 1972) has proposed that both the positive cooperativity expressed by cyclic GMP in the catalysis of cyclic AMP as well as the negative cooperativity expressed by low substrate concentrations of cyclic AMP for its own catalysis may have functional significance: formation of cyclic GMP could activate the destruction of an opposing cyclic AMP (see GOLDBERG et al., 1973a, b) while small amounts of cyclic AMP could inhibit its own breakdown at least long enough to result in the activation of the appropriate protein kinase.

3. Possible Principles of Cyclic Nucleotide-Sensitive Enzyme Reactions

Several recent publications have emphasized the presence of calcium or other cation-dependent proteins which can regulate (usually by activating) the activity of adenylate cyclase (BROSTROM et al., 1975), phosphodiesterases (KAKIUCHI et al., 1975; WICKSON et al., 1975; UZUNOV et al., 1975) and phosphoprotein phosphatase (MAENO et al., 1975). In the latter report, GREENGARD and his colleagues suggest three possible mechanisms by which elevations of intracytoplasmic cyclic AMP levels could result in the activation of the enzyme which dephosphorylates the synaptic membrane protein substrate, Protein II: (1) cyclic AMP directly activates the phosphoprotein phosphatase; (2) cyclic AMP activates a protein kinase which activates the phosphatase; (3) cyclic AMP is able to increase accessibility of phosphorylated Protein II substrate to the phosphatase.

Since the conditions for assay of Protein II dephosphorylation required the total inhibition of all protein kinase activity (with calcium chelators and adenosine) possibility 2 seems quite unlikely. On the other hand, possibility 3 is favored by the similarity between the molecular properties of Protein II and protein substrates in other tissues, such as heart, which can bind to protein kinase and inhibit the enzyme, but which, in the dephosphorylated form, disassociates from the enzyme, resulting in activation. In the present case, activation by cyclic AMP of a Protein II kinase to phosphorylate Protein II would presumably lead to transient association of Protein II with the kinase to become phosphorylated. Subsequent dissociation of Protein II from the kinase would permit the substrate to be dephosphorylated by the already active phosphatase. This overall effect would result in a temporary displacement of Protein II, similar to a Sol-Gel transformation, which would presumably permit some subsequent molecular event to be initiated (or terminated), but it remains unclear whether the crucial event is the interaction on Protein II (or some other substrate) or on the larger proteins with which they can become associated.

If these types of interactions also pertain to other related enzymes, a Ca-sensitive protein which is transiently phosphorylated could be a common regulatory principle: a protein normally bound to a nucleotide cyclase. Upon dissociation of this regulator the cyclase is activated and then the free regulator could then be dephosphorylated, in which form it activates a cyclic nucleotide phosphodiesterase (BROSTROM et al., 1975; WICKSON et al., 1975).

E. Summary: Some Criteria for Establishing that Second Messenger Systems Mediate Responses to Hormones or Neurotransmitters

In Section II.C, the criteria were briefly considered by which the chemical identity of a transmitter for a specific neuronal pathway can be established. How might these criteria now be modified to apply to the question of whether the particular pathway or transmitter affects the postsynaptic cell by a cyclic nucleotide system?

This question might be approached by paraphrasing the four criteria developed by SUTHERLAND (SUTHERLAND et al., 1965; ROBISON et al., 1971) for establishing that the action of any given hormone was mediated by cyclic AMP:

1. The Neurotransmitter Substance and the Activation of the Synaptic Pathway will Regulate the Intracellular Levels of Cyclic Nucleotide in the Postsynaptic Cell Population. With few exceptions however, (see IV.C1) the methods for detection of cyclic nucleotides cannot resolve them within specific cells, and therefore a variety of lower resolution methods have been employed to evaluate the effects of electrical stimulation or neurotransmitters in intact ganglia, in intact cells within brain slices (Table 1) or to determine the effects of transmitter substances in homogenates of brain (Table 2).

2. The Change in Intracellular Cyclic Nucleotide Content Should Precede the Biological Event Triggered by the Transmitter or Nerve Pathway. This criterion is also difficult to satisfy since the sampling methods of cyclic nucleotide measurement have so far not been sufficiently rapid to detect changes within the period of time in which synaptic potentials are generated. Repetitive activation of a pathway could be used to produce a measurable change even though it may also result in non-physiological changes uncharacteristic of the normal mode of a pathway's actions. Furthermore, this criterion demands that a specific biological event (eg., inhibition, excitation) for the action of a given pathway be demonstrable; these events remain undefined even in simple terms for most brain pathways. While the timing of a change in cyclic nucleotides relative to a synaptic potential is crucial for cause and effect demonstrations, it is conceivable that experimental changes sufficiently marked to yield measurable changes in nucleotide levels may so saturate the systems for catalysis that nucleotide levels remain abnormally high for durations longer than the short-lived acute synaptic potential.

3. The Effect of the Transmitter or Nerve Pathway in Eliciting the Physiological Event Should be Altered by Drugs which can Specifically Prevent the Hormonal Response of the Nucleotide Cyclase or which can Inhibit the Appropriate Phosphodiesterase. In view of the current inability to satisfy the first two criteria within the proper temporal and spatial resolutions required to study discrete brain pathways, heavy emphasis must now be placed upon acquiring pharmacological evidence. Thus, potentiation of the effects of a synaptic pathway by phosphodiesterase inhibition or blockade of the effects of a pathway by drugs which block the hormonal activation of the cyclase becomes critical. However, when this approach is taken, it must be done with the proviso that each drug interaction must be verified for specificity of action as far as is possible.

4. Exogenous Cyclic Nucleotides Should be Able to Elicit the Biological Event Caused by the Transmitter or Nerve Pathway. However, living cells are relatively impermeable to organic phosphates such as cyclic nucleotides (KAUKEL and HILZ, 1972). Since nucleotides applied to the exterior of the cell are not only distant from their presumed natural site of production but exposed to the catalytic action of soluble phosphodiesterases as well, this criterion also carries considerable

Table 1. Changes in cyclic AMP dynamics in nervous tissue *in vitro*. This table indicates the species and regional variations in the effects of neurotransmitters on cyclic nucleotide concentrations or synthesis rates in tissue slices. Because of qualitative variations between individual reports, results are reported in ranges of multiples of the control value; the variations expressed by these ranges requires that all of the papers for a given species and region be consulted for further evaluation (also see RALL and SATTIN, 1970; RALL, 1971; VON HUNGEN and ROBERTS, 1974; DALY, 1973, 1975)

Nervous tissue source	Agent[a]	Effect[b]	Interactions[c]	References
I. Invertebrates				
A. Aplysia ganglia	Elec. Stim.	2×(L)	b: 0.2M Mg^{++}	CEDAR et al., 1972
	DA 200	3×(L); 6×(A)		
	5-HT 200	3×(L); 8.5×(A)		CEDAR and SCHWARTZ, 1972
	Octopamine 100	3–5(A)		LEVITAN et al., 1974
B. Cockroach ganglia	Octopamine 25	7×(L)		NATHANSON and GREENGARD, 1973
II. Frog				
A. Sympathetic ganglia	Elec. Stim.	7.8×(L)		WEIGHT et al., 1974
B. Brain	EPI 100	1.1×(L)		FORN and KRISHNA, 1971
III. Mouse				
A. Sympathetic ganglia				
1. Cultured neuroblastoma	PGE-1 2.6	10×(L)		GILMAN and NIRENBERG, 1971a
	DA 10	2.0×(S)	b: haloperidol	PRASAD and GILMER, 1974
	NE 100	1.2×(S)	b: propranolol	PRASAD et al., 1973
	ISO 100	2.0×(S)		PRASAD et al., 1973
			p: Ro-20-1724	BLUME et al., 1973
	ADEN 200	2.5×(L)	p: Ro-20-1724	BLUME et al., 1973
2. "Differentiated" neuroblastoma	DA 10	2.1×(S)		PRASAD and GILMER, 1974
	NE 10	1.4×(S)		PRASAD and GILMER, 1974
	ISO 100	0		PRASAD and GILMER, 1974
B. Brain				
1. Reaggregate cultures	NE, ISO 100	5.0×(L)		SEEDS and GILMAN, 1971
2. Cerebral cortex	NE, ISO 100	4–8×(L)(A)	b: propranolol	RALL and SATTIN, 1970
	HIST 100	1.6×(A)		FORN and KRISHNA, 1971; SCHULTZ and DALY, 1973
	ADEN 100	15×(A)	(decreased in quaking mutant)	SKOLNICK and DALY, 1974
3. Cerebellum	GLU 10000	10×(L)	p: NE b: EGTA, low Ca	FERRENDELLI et al., 1974

[a b c] See p. 23.

Table 1 (continued)

Nervous tissue source	Agent[a]	Effect[b]	Interactions[c]	References
IV. Rat				
A. Sympathetic ganglia (whole cultures)	EPI 100	16×(L)	b: propranolol	CRAMER et al., 1973
	NE 1000	14×(L)	b: propranolol	CRAMER et al., 1973
	ISO 1.0	25×(L)	b: propranolol	CRAMER et al., 1973; OTTEN et al., 1973
	DA 10000	1.5×(L)		CRAMER et al., 1973
	HIST 100	2×(L, A)	b: diphenhydramine	LINDL and CRAMER, 1974
B. Brain				
1. Monolayer cultures whole brain	ISO 1.0	94×(L)	with theophylline	GILMAN and SCHRIER, 1972
	NE 1.0	17×(L)	with theophylline	GILMAN and SCHRIER, 1972
	ADEN 100	12×(L)		
	PGE-1 3.0	30×(L)		GILMAN and SCHRIER, 1972
2. Glioma culture	ISO 6.6	10×(S)	p: Li, Na, K	SCHIMMER, 1971
	EPI 6.6	6×(S)		
	NE 6.6	4.5×(S)		
	NE 100	90–1000×(A)		SCHULTZ et al., 1972
	NE 3.0	5×(L)	refractoriness b: cycloheximide	DEVELLIS and BROOKER, 1974
3. Cerebral cortex	NE 30-100	4–6×(L) 0.3–10×(A)	b: Li; protryptiline p: phosphodiesterase inhibitors	RALL and SATTIN, 1970; FORN and KRISHNA, 1971; FORN and VALDECASSAS, 1971; PERKINS and MOORE, 1973; HUANG et al., 1973; PALMER, 1973; PALMER and SCOTT, 1974; SKOLNICK and DALY, 1974
			p: 6-OHDA pretreatment	KALISKER et al., 1973; HUANG et al., 1973; PALMER and SCOTT, 1974
	ISO, EPI 100	2–3×(L)	b: partial with either phentolamine or propranolol	SCHULTZ and DALY, 1973c
	ADEN 100	4–5×(L)	p: catecholamines, HIST and 5-HT	SCHULTZ and DALY, 1973c
4. Cerebellum	NE 100	2×(A) 5×(L)		RALL and SATTIN, 1970; FORN and KRISHNA, 1971; PALMER et al., 1973
	ADEN 100	1.2×(L)		RALL and SATTIN, 1970
5. Hypothalamus	NE 50	3.5×(L)	b: Li; chlorpromazine	PALMER et al., 1973
6. Estradiol		2–3×(L)	b: alpha adrenergic	GUNAGA and MENON, 1973
7. Hippocampus	NE 50-100	3–5×(L, A)	b: LSD	FORN and KRISHNA, 1971; PALMER and BURKS, 1971; PALMER, 1973; PALMER et al., 1973

[a] [b] [c] See p. 23.

Table 1 (continued)

Nervous tissue source	Agent[a]	Effect[b]	Interactions[c]	References
8. Caudate nucleus	DA 100 NE 100	2×(L) 3.5×(L)	b: fluphenazine b: propranolol	FORN et al., 1974 FORN et al., 1974
V. Guinea Pig				
A. Brain				
1. Cerebral cortex	EPI, NE 10-100	1.2–4(A)(L)	p: HIST b: phentolamine	KAKIUCHI et al., 1969; SHIMIZU et al., 1969; SATTIN and RALL, 1970; FORN and KRISHNA, 1971; HUANG et al., 1971
	HIST 10-100	1.5–20.0(A)(L)	p: EPI: b: diphenhydramine, chlorpheniramine	Same as EPI, NE; CHASIN et al., 1973; SCHULTZ and DALY, 1973a–c; BAUDRY et al., 1975
	ADEN 100	15–50×(A) 15–50×(L)	p: HIST, 5-HT b: partial theophylline	SATTIN and RALL, 1970; RALL and SATTIN, 1970; SCHULTZ and DALY, 1973a
	GLU, ASP 1000–10000	40–80×(A)	p: ADEN, HIST	FERENDELLI et al., 1974; SHIMIZU et al., 1974
	Elec. Stim.	11–20×(L)	a: NE, HIST b: anesthetics	KAKIUCHI et al., 1969 ZANELLA and RALL, 1973
2. Cerebellum	NE, ISO 100	8–20×(L)(A)	b: propranolol	CHASIN et al., 1971; RALL, 1971; ZANELLA and RALL, 1973
	ADEN 100	50×(L)(A)	b: theophylline partial b: K^+	SATTIN and RALL, 1970; RALL and SATTIN, 1970; ZANELLA and RALL, 1973
3. Hippocampus, amygdala	EPI 100	1.5×(L)(A)	b: partial phentolamine, dibenamine, phenoxybenzamine p: HIST	CHASIN et al., 1973
	HIST 10	5.5×(L)(A)	b: chlorpheniramine; p: EPI	CHASIN et al., 1973
VI. Rabbit				
A. Sympathetic ganglion	Elec. Stim.	5.0×(L)	b: atropine and hexamethonium b: atropine	MCAFEE et al., 1971 KALIX et al., 1974
B. Brain				
1. Cerebral cortex	NE 100	1.5–2.0(L)(A)		KAKIUCHI and RALL, 1968; SHIMIZU et al., 1969; FORN and KRISHNA, 1971; SCHMIDT and ROBISON, 1971
			b: Li	FORN and VALDECASAS, 1971
	HIST 100	3–18×(L)(A)	p: ADEN b: diphendydramine	Same as NE; BERTI et al. 1972; KUO et al., 1972; PALMER et al., 1972

[a] [b] [c] See p. 23.

Table 1 (continued)

Nervous tissue source	Agent[a]	Effect[b]	Interactions[c]	References
2. Cerebellum	NE 100	5–16×(L)(A)	p:ADEN b:dichloro- isoproterenol	KAKIUCHI and RALL, 1968; SHIMIZU et al., 1969, 1970; FORN and KRISHNA, 1971; BERTI et al., 1972 KUO et al., 1972
	HIST 100	3–10×(L)(A)		
3. Hippocampus, hypothalamus	NE 100 HIST	1–2(L)		SCHMIDT and ROBISON, 1973
VII. Cat				
A. Sympathetic ganglion	Elec. Stim.	1.6×(A)		CHATZKEL et al., 1974
B. Brain				
1. Cerebral cortex, cerebellum	NE 100	3–5×(A)		FORN and KRISHNA, 1971
2. Caudate nucleus	HIST 100	1.5–2.0(A)		
VIII. Cow				
A. Sympathetic ganglia	NE, DA 100	5–7×(L)		KEBABIAN and GREENGARD, 1971
B. Retina	DA 100	1.8×(A)	b:fluphenazine haloperidol	BROWN and MAKMAN, 1971, 1973
IX. Monkey				
A. Cerebral cortex (squirrel monkey)	NE 100 5-HT 100	4–5×(A) 1–2×(A)	p:ADEN	SKOLNICK et al., 1973
B. Cerebellum (rhesus)	NE 100	1–2×(A)		FORN and KRISHNA, 1971
C. Hypothalamus (″)	NE 100	5.5×(A)		FORN and KRISHNA, 1971
X. Human				
A. Astrocytoma cultures	NE 30	5×(L)		CLARK and PERKINS, 1971
	HIST 30	8×(L)		CLARK and PERKINS, 1971
B. Cerebral cortex, cerebellum	NE 100 HIST 100 ADEN 100	12–20×(A) 0.2×(A) 2–6×(A)		SHIMIZU et al., 1971; BERTI et al., 1972; FUMAGILLI et al., 1971; KODAMA et al., 1973

Key: The major studies in which neurotransmitters, adenosine, or electrical stimulation, were applied to intact cell preparations of central or peripheral nervous tissue are grouped according to species and brain region.

[a] Agents in micromoles: NE, norepinephrine; DA, dopamine; EPI, epinephrine; ISO, isoproterenol; ADEN, adenosine; 5-HT, serotonin; GLU, glutamate; ASP, aspartate.

[b] Effect: the range in multiples of control values by which cyclic AMP was increased in terms of levels (L), or the accumulation of labeled cyclic AMP after exposure to labeled adenosine (A) or labeled ATP (S).

[c] Interactions: the relative influence of other drugs: a: additive changes; b: effect blocked; p: effect potentiated.

Table 2. Effects of neurotransmitters and adenosine on the synthesis of cyclic AMP by cell-free homogenates of nervous tissue. All abbreviations and notes as in Table 1

Nervous tissue source	Agent[a]	Effect[b]	Interactions[c]	References
I. Invertebrate				
Cockroach ganglia	Octopamine 30	4 × (S)	b: phentolamine	NATHANSON and GREENGARD, 1973
	DA 30	2.5 × (S)	a: octopamine; 5-HT	NATHANSON and GREENGARD, 1973
	5-HT 25	1.7 × (S)	b: LSD, Br-LSD; cyproheptadine	NATHANSON and GREENGARD, 1974
II. Frog				
Retinal rod Outer segments	Photic Stimulation	decreased (L)	ATP dependent-phosphodiesterase activation	MIKI et al., 1973; BITENSKY et al., 1973
III. Mouse				
Caudate nucleus	DA 100	2.1 × (S)		CLEMENT-CORMIER et al., 1974
IV. Rat				
Whole retina	DA, NE 200	0.4 × (S)		BROWN and MAKMAN, 1972
Cerebral cortex	NE 10–100 EPI 100	1.0–1.2 × (S) 1.2 (L)	b: propranolol	JANIEC et al., 1970; MCCUNE et al., 1971; VON HUNGEN and ROBERTS, 1974; WALKER and WALKER, 1973a; KATZ and TENEHOUSE, 1973
Cerebellum	NE 50	1.8 × (S)		WALKER and WALKER, 1973b
Caudate Nuc/Limbic Forebrain	DA 20–100	2.0 × (S)	b: phentolamine, and anti-psychotics	KEBABIAN et al., 1972; CLEMENT-CORMIER et al., 1974; WALKER and WALKER, 1973b; KAROBATH and LEITICH, 1974; MISHRA et al., 1974; SHEPPARD and BURGHARDT, 1974; MILLER et al., 1974
V. Guinea Pig				
Cerebral cortex	EPI 10-100 HIST 100	1.8 × (A) (L) 3.0 × (A) (L)	b: dibenamine	CHASIN et al., 1974
Cerebellum	EPI 10	6 × (A) (L)	b: propranolol	CHASIN et al., 1974
Hippocampus	HIST 10	3–4 × (A) (L)	b: chlorphenir-amine; p: ADEN; cyclic AMP	CHASIN et al., 1974
Caudate nucleus	DA 100	2 × (S)		CLEMENT-CORMIER et al., 1974
VI. Rabbit				
Caudate nucleus	DA 100	2 × (S)		CLEMENT-CORMIER et al., 1974
VII. Cat, sheep				
Cerebellum	NE, ISO, EPI 100	1–3 × (L)		KLAINER et al., 1962

[a b c] See p. 23.

difficulties for objective satisfaction. Generally, large amounts of exogenous nucleotide with phosphodiesterase inhibitors are required to produce hormone-like effects. A large spectrum of synthetic nucleotides has been prepared (POSTERNAK et al., 1962; SIMON et al., 1973; MILLER et al., 1973; MEYER and MILLER, 1974; TSOU et al., 1974) with specific improvements in lipo-permeability and resistance to phosphodiesterase action so that this criterion can now be met with less difficulty.

5. To this list of Sutherland's original criteria may now be added the proposal by KUO and GREENGARD (1969a, b), that *activation of a phosphotransferase reaction is the major mechanism by which alterations of cyclic nucleotide levels are expressed.* To apply this criterion, for example, *the action of a synthetic cyclic nucleotide ought to emulate the effects of the transmitter or pathway on the parameter of the biological event in direct correlation to the ability of that nucleotide to activate the appropriate protein kinase reaction.* In fact, a separate list of criteria has been proposed for establishing that activation of a protein kinase is the mechanism for expression of a hormone-activated cyclase-coupled response (see LANGAN, 1972), but these depend mainly upon the ability to discern in the appropriate substrate protein a functional relationship to the biological response. Such a relationship currently would seem difficult to achieve for the question of interneuronal transmission, until the substrate proteins can be better characterized functionally. Therefore, this fifth criterion can only be applied now to statistical correlations based on general kinase activation constants (see DRUMMOND and POWELL, 1970; SIMON et al., 1973; SWILLENS et al., 1974; KUO et al., 1974).

These criteria establish a common yardstick against which to measure the degree of progress in the analysis of the action of a given transmitter at a given synaptic connection. Although it is currently difficult to satisfy all of the criteria, evidence (to be reviewed below) suggests that several specific examples have now been described in which very strong cases have been built for the mediation of synaptic transmission by cyclic AMP through the use of multiple lines of mutually supportive observation.

IV. The Assessment of Neurobiologic Events Which may be Mediated by Cyclic Nucleotides

In order to proceed to analyze the possible involvement of cyclic nucleotide chain reactions in the expression of the response of a nervous tissue cell to neurotransmitters or to more general hormones, some critical awareness of the powers and limitations of the various technological approaches involved may be helpful. Two major methods have been employed. (1) *Biochemical estimations* in which the effects of hormones or transmitter substances are evaluated against changes in cyclic nucleotide dynamics or in the effects of the cyclic nucleotides upon phosphotransferase reactions. (2) *Electrophysiological estimates,* in which the biophysical and pharmacological response of neurons or other innervated target cells to nerve pathways or exogenous neurotransmitters, are used to gauge the effects of exogenous cyclic nucleotides or drugs which affect cyclic nucleotide dynamics. Clearly, these two major approaches are interdependent and symbiotic, and the following sections seek to present some of the factors required for critical evaluation of the data which can be generated from these approaches, as well as from an intermediate approach less vigorously pursued at present, the *cytochemical localization* of the cyclic nucleotides and their related enzyme activities.

A. Biochemical Assessment

The recent extensive reviews by PERKINS (1973), LANGAN (1972), and DALY (1975) provide extensive treatments of the potential pitfalls and limitations of the biochemical estimation of cyclic nucleotide levels and cyclic nucleotide synthesis, degradation and subsequent action on biochemical processes. Moreover, the entire second volume of Advances in Cyclic Nucleotide Research (1972) was devoted to an updating of the methods for the chemical measurement of the nucleotides.

For purposes of evaluating the functional consequences of cyclic nucleotide dynamics in nervous tissue, several issues deserve additional discussion. At one end of an analytical spectrum, the brain could be regarded as a homogeneous organ regulating behavior, such that some functional insight might derive from changes in behavior which accompany the response to a drug influencing whole brain cyclic nucleotides. However, administration of drugs in sufficient amounts to influence either behavior or cyclic nucleotide levels requires a time span of drug administration, chemical sampling and analysis which is orders of magnitude beyond the scale of individual synaptic events. Therefore, while such psychopharmacological interactions can be sought and quantified, interpreting the cellular or synaptic basis of these data is far more difficult. At the opposite end of the spectrum would be evaluation of the molecular effects of transmitters on cyclic nucleotides or phosphotransferase reactions within specific populations of potential target cells. While partially approached by studies of cultured neural cell lines (see PERKINS, 1973; GILMAN, 1972), it has still not been possible to obtain sufficient amounts of homogeneous differentiated, intact, target cells free of surrounding cellular material to pursue such molecular changes substantially (however, see GREENGARD and KEBABIAN, 1974). In between these extremes, the greatest amount of biochemical data has come from preparations in which brain slices are challenged over many minutes with agents which may influence cyclic nucleotide dynamics (Table 1); in a growing number of more recent cases, conditions have been found in which cell free preparations of nervous tissue can be explored for the same purposes (Table 2).

The brain slice preparation (see KAKIUCHI and RALL, 1968a, b; KAKIUCHI et al., 1969; RALL and SATTIN, 1970; DALY et al., 1972; DALY, 1975; SATTIN et al., 1975) has come to represent a convenient intermediate situation in which brain can be submitted to more controlled environments than within the rapidly changing brain in vivo, and which still exhibits sufficient responsiveness to transmitters and related perturbations to advance the state of the art. The cells are perhaps less "intact" and possibly somewhat sub-viable compared either to brain in situ or to re-aggregated or monolayer nerve tissue cultures, but their improved ease of handling provides tremendous relative practical benefits. More troublesome aspects still eluding critical evaluation are the hetereogeneity of the cells within a brain slice, (both in terms of their heterogeneity to each other and to their presumably homologous cells in adjacent slices), the relative response of different cell types within the slice to the in vitro situation, and the possible dilution of small but specific dynamic changes in some hormone target cells by non-responsive cells (see RALL, 1972; RALL and SATTIN, 1970).

In general, the changes in cyclic nucleotides which occur in brain slices have been evaluated in terms of the absolute level of nucleotide per unit tissue, or in terms of the relative synthesis of cyclic AMP from an isotopically labeled precursor, usually adenine or adenosine (see DALY, 1975 for an extensive discussion). The latter index is usually referred to as "accumulation" rather than synthesis; synthesis can be assessed directly by conversion of labelled ATP to labelled cyclic AMP, but this is usually reserved for cell free systems to avoid problems with intracellular precursor pools.

However, the levels of cyclic nucleotides in the brain (like other rapidly synthesized metabolic products, see LOWRY et al., 1964) show rather marked alterations simply upon excision of the brain (BRECKENRIDGE, 1964; WEISS and COSTA, 1968; also see STEINER et al., 1972c; KIMURA et al., 1974). Because of this problem, rather innovative methods have been employed to remove brain samples which would reflect the levels as they were during the living state. For this purpose, modifications of the microwave oven (STAVINOHA et al., 1970) have been adapted for determinations of brain cyclic nucleotides (SCHMIDT et al., 1971). This method can "fix" brain rapidly enough so that levels of oxidative metabolites measured after this procedure are quite close to the values obtained when the brain is "freeze-blown" (LUST et al., 1973; also see GUIDOTTI et al., 1974). Since the overall brain structure remains relatively intact after micro-wave irradiation, small pieces can still be isolated for regional analyses, although all enzymatic activity is, of course, destroyed in the process.

The brain slice, on the other hand, permits evaluation of a more dynamic interaction between transmitters and enzyme activity, but the determination of the levels of nucleotides within slices in vitro requires periods of stabilization to permit recovery from the agonal changes. Furthermore, adenosine can be a potent stimulant of adenylate cyclase in most brain slice preparations (see below), and several other depolarizing agents such as ouabain, batrachatoxin, veratridine high K^+ (DALY et al., 1972; DALY, 1975) or electrical stimulation (see ZANELLA and RALL, 1973) may exert much of their effect through the release of adenosine within the slice. Therefore, since adenosine release increases under conditions of anoxia (see RALL and SATTIN, 1970; SATTIN and RALL, 1970; BERNE et al., 1974; DALY, 1975; SATTIN et al., 1975), the basal levels and relative changes after transmitter or drug exposure must in some way control for this alternate source of change. Nevertheless, despite these problems in the interpretation of brain slice data, this general approach has provided the majority of information now available on the ability of transmitter substances to increase the tissue concentration of cyclic AMP or cyclic GMP.

B. Neurotransmitters Affecting Tissue Levels of Cyclic Nucleotides

A very detailed resumé of the extremely large number of studies employing brain slices to evaluate the ability of neurotransmitter substances to influence the levels of cyclic nucleotides has recently been presented by DALY (1975; also see RALL, 1972; RALL and GILMAN, 1970; DALY et al., 1972). These studies are also recapitulated according to species and nervous tissue region in Table 1; the effects of transmitters in cell-free preparations of nervous tissue are outlined in Table 2. Due to the extensive literature contained in these tables, this section will summarize the data from a few prototypic reports. Analysis of these data indicate that three groups of substances produce the most pronounced effects: the catecholamines, epinephrine, norepinephrine, or dopamine; the biogenic diamine, histamine; and adenosine. However, the ability of these and other factors, such as serotonin or octopamine, to increase cyclic AMP levels cannot be generalized across species or brain regions and each area of each species seems to have its own individual pattern of responsiveness. Inexplicably, some substances capable of altering hormonal responses in other tissues, such as the prostaglandins (see ROBISON et al., 1971; SIGGINS et al., 1971c) have little effect in slices (ZANELLA and RALL, 1973). In this section, the major substances effective in altering cyclic nucleotide levels are reviewed for their actions in various species and regions.

1. Catecholamines

In most brain slice preparations, the most effective catecholamine in increasing accumulation of cyclic AMP is norepinephrine; increases of 1.5–10 fold or more are observed in the cerebral cortex of all mammals, with the lowest stimulation being seen in rabbit and guinea pig cortex and some of the greatest in samples of human cortex. In preparations composed of cerebellar slices, norepinephrine also produces increases of 2–20 fold, with the rat being at the lower end of the incremental scale, perhaps due to a higher basal activity under these conditions. Norepinephrine has also produced increases of less than one order of magnitude in slice preparations of limbic system, hypothalamus and mesencephalon. Pharmacologically, the receptor involved in these responses to norepinephrine is characterized as a beta receptor in the cerebellum; but in the other regions, both alpha and beta blockers are partially antagonistic. A small increase in cyclic AMP-content in response to norepinephrine has also been observed in slices of the rat caudate nucleus (WALKER and WALKER, 1973a; FORN et al., 1974), and this response was blocked by propranolol and partially blocked by the antipsychotic fluphenazine (FORN et al., 1974). In slices of rat limbic forebrain, BLUMBERG and SULSER (1974) reported preliminary results that norepinephrine, but not dopamine, increases cyclic AMP accumulation 4–7 fold, and that micromolar concentrations of several antipsychotic drugs antagonized the norepinephrine effect. These latter pharmacological observations stand in distinction to the large number of reports describing the effectiveness of antipsychotics in blocking the action of dopamine on adenylate cyclase (see below).

Dopamine has not as yet been shown to increase cyclic AMP levels or conversion of adenosine or adenine to cyclic AMP in *slices* of any brain region except the caudate nucleus (FORN et al., 1974; but see WALKER and WALKER, 1973a). However, dopamine will increase cyclic AMP (absolute levels and the synthesis of cyclic AMP from labeled ATP) in cell-free preparations of caudate nucleus of several species (KEBABIAN and GREENGARD, 1971; KEBABIAN et al., 1972; CLEMENT-CORMIER et al., 1974; MILLER et al., 1974; SHEPPARD and BURGHARDT, 1974), as well as in homogenates and whole retina (BROWN and MAKMAN, 1972, 1973) and in slices of bovine sympathetic ganglia (KEBABIAN and GREENGARD, 1971). Dopamine does not stimulate cyclic AMP formation in cultured rat ganglion (CRAMER et al., 1973) or in the intact rabbit ganglion (KALIX et al., 1974). In the cell-free preparations of caudate nucleus and other dopamine-rich regions of the forebrain, the effects of dopamine are generally greater than those of norepinephrine and the dopamine effect is extremely sensitive to sub-micromolar concentrations of most, but not all antipsychotics (see Table 2; also V.C). In the ganglionic nervous systems of invertebrates, the analogous phenylethylamine transmitter substance would appear to be octopamine: the beta-hydroxylated, decarboxylated product of tyrosine. In cockroach (NATHANSON and GREENGARD, 1973) and in Aplysia (LEVITAN et al., 1974), octopamine produces a marked rise in cyclic AMP synthesis, blocked by phentolamine, an alpha antagonist; dopamine can also increase the accumulation of labeled cyclic AMP in Aplysia (CEDAR and SCHWARTZ, 1972).

The increase in cyclic AMP levels observed in response to biogenic amines can usually only be triggered once in a given incubation of brain slices (see RALL and SATTIN, 1970; DALY, 1975). This loss of effectiveness, which does not occur for adenosine, may be explained by the recent report by DEVELLIS and BROOKER (1974) that in a rat glioma cell line, cycloheximide prevents refractoriness of cyclic AMP synthesis stimulation by norepinephrine (also see SCHULTZ et al., 1972). The inference here is that part of the original hormone-induced response is the induction or activation of a phosphodiesterase (see UZUNOV et al., 1973).

These biochemical effects of the catecholamines, summarized above, provide good starting points for electrophysiological tests of the synaptic functions of cyclic AMP (see below, IV.C) since the anatomical projections of these catecholamine pathways can now be traced from source to target by several cytochemical methods (see UNGERSTEDT, 1971; HÖKFELT and UNGERSTEDT, 1972, 1973; BLOOM, 1973).

2. Histamine

Histamine is the other major transmitter substance effective in increasing the accumulation of cyclic AMP from labeled adenosine or adenine in several brain regions. Recently, Schwartz and collaborators (GARBARG et al., 1974) have presented the first convincing demonstration that cortical histamine is contained in specific nerve tracts: histamine concentration was found to decrease after electrolytic destruction of the medial forebrain bundle. These data extend previous results form several groups which indicated that histamine content was greater in subcellular fractions enriched in synaptic terminals (cf., GARBARG et al., 1974). Thus it is likely, although not yet established, that changes in cyclic AMP levels in response to histamine could carry the same implications for synaptic transmission as the responses described above for catecholamines.

The major responses to histamine are observed in the rabbit cerebellum (KAKIUCHI et al., 1969; SHIMIZU et al., 1970a, b, c; FORN and KRISHNA, 1970, 1971; also see Table 1) and in the cerebral cortex and hippocampus of the guinea pig (PALMER et al., 1972b; CHASIN et al., 1973, 1974; BAUDRY et al., 1975). Histamine can also stimulate cyclic AMP accumulation in the rat sympathetic ganglion (LINDL and CRAMER, 1974), but the effect is substantially smaller than the response of the ganglion to norepinephrine. The histamine effects are usually blocked at least partially by antihistaminics of either the H_1 or H_2 receptor classes (BAUDRY et al., 1975) and therefore histamine does not seem to be acting as a partial agonist acting on receptors for catecholamines. Histamine effects are in general potentiated greatly by adenosine.

3. Serotonin

The other major biogenic monoamine of the mammalian nervous system is serotonin, and this substance has also been reported to produce some increase in cyclic AMP accumulation in the guinea pig and rat cerebral cortex; however this effect is minimal when the slices are exposed to serotonin as the sole agonist, and are seen mainly as a potentiated response to a combination of serotonin and adenosine (SCHULTZ and DALY, 1973c; DALY, 1975). In the cockroach ganglion cell-free system, NATHANSON and GREENGARD (1974) observed direct actions of serotonin on cyclic AMP synthesis and reported that this action of serotonin could be blocked by very low amounts of LSD, Bromo LSD, and Cyproheptadine (but see AGHAJANIAN et al., 1972 for a differing view of the mammalian brain LSD action). Serotonin also elevated the cyclic AMP accumulation rate of intact Aplysia abdominal ganglia (CEDAR and SCHWARTZ, 1972).

4. Acetylcholine

At present, there have been relatively few reports in which acetylcholine or cholinesterase-resistant derivatives of acetylcholine were able to alter cyclic nucleotide levels in vitro. However, in slices of whole bovine or rat cerebrum (LEE et al., 1972) and in slices of bovine sympathetic ganglia (see GREENGARD and KEBABIAN, 1974), acetylcholine has been shown to increase the content of cyclic GMP. These reports and studies of stimulated frog sympathetic ganglia (WEIGHT et al., 1974) are compatible with the idea that a muscarinic cholinergic receptor is responsible for the elevations of cyclic GMP content. There have also been several studies of parenterally administered muscarinic agonists (FERRENDELLI et al., 1970, 1972, 1973) which elevate regional levels of cyclic GMP in mouse brains quick-frozen in vivo, but lack of characterization of these receptors makes interpretation difficult.

5. Adenosine, Amino Acids, and Polypeptides

Several comprehensive research reports have characterized the manner in which adenosine leads to enhanced accumulation of cyclic AMP in tissue slices (RALL and SATTIN, 1970; SATTIN and RALL, 1970; ZANELLA and RALL, 1973; SCHULTZ and DALY, 1973a, b, c; DALY, 1975; SATTIN et al., 1975). These studies indicate that adenosine can, in some cases, produce cyclic AMP rises which are greater than additive when administered with one of the amine transmitter substances or with some other form of chemical or electrical depolarization. However, other evidence suggests that the effects of adenosine are mediated through receptors which are distinct from the transmitter receptors. One striking property of the adenosine receptors is that they are blocked by methyl xanthine phosphodiesterase inhibitors, such as theophylline, which generally potentiate the effects of monoamines. All other adenine nucleotides which can elevate cyclic AMP accumulation in slices, such as ATP, ADP or 5′-AMP, are thought to depend for this effect upon their hydrolysis to adenosine (see SATTIN and RALL, 1970; RALL and SATTIN, 1970; RALL, 1971; DALY, 1975).

The earliest studies on the effects of adenosine derived from incubated cerebral slices (SATTIN and RALL, 1970). Unlike the potentiative effects of methyl xanthine phosphodiesterase inhibitors (see below) in the elevation of cyclic AMP levels by transmitter substances, the stimulation of cyclic AMP accumulation by adenosine was inhibited by methyl xanthines (SATTIN and RALL, 1970). While subsequent studies indicated that stimulating doses of adenosine did increase the pool of cyclic AMP precursor adenine nucleotides (see DALY, 1975 for review), it also appeared that adenosine was able to increase cyclase activity although not directly. Additional studies indicated potentiative interactions between adenosine and histamine or norepinephrine (see RALL, 1971; SATTIN et al., 1975; DALY, 1975) and that (at least in the rat and guinea pig cerebral cortex), an alpha adrenergic receptor activated adenylate cyclase only in the presence of adenosine (SCHULTZ and DALY, 1973c, d). These observations further extended the view (see DALY et al., 1972) that adenosine could itself initiate changes in cyclic AMP from an extracellular receptor. In human astrocytoma cells (CLARK et al., 1974), the likelihood of an extracellular site of action for adenosine and the adenine nucleotides, AMP, ADP and ATP, is strengethened by the observation that dipyridamol—which blocks adenosine accumulation (see DALY, 1975)—does not depress the extent of the cyclic AMP accumulation induced by adenosine. Most recently, SATTIN, RALL and ZANELLA (1975) have proposed the interesting concept of adenylate cyclase co-agonists in which adenosine and a catecholamine or histamine could act either as dependent or independent co-activators of the cyclase. If the ability of adenosine nucleotides to alter cell firing (GALINDO et al., 1967; PHILLIS et al., 1975) were related to such adenosine receptors, then it may be appropriate to consider extending into the central nervous system the concept of purinergic transmission which BURNSTOCK has proposed for the toad intestine (1972). It is of interest that SATTIN and RALL (1970) also suggested that ATP (which is stored with catecholamine in presynaptic terminals) might also be secreted with the catecholamines and thereby supplement their effects in activating the postsynaptic cyclase. Subsequent experiments strongly support their foresighted prediction, and suggest that many substances may be able to influence neuronal function and metabolism through changes in intracellular cyclic nucleotides.

Recent studies (FERRENDELLI et al., 1974; SHIMIZU et al., 1974) indicate that some of the effects of adenosine in brain slices could be mediated by the release of excitatory amino acids, namely those with di-acidic functions such as glutamate or aspartate (see CURTIS and JOHNSTON, 1974). However, the concentration of amino acid required to produce elevations of cyclic nucleotides is, in general, 2–3 orders of magnitude higher than the concentration of the monoamine transmitters necessary to produce comparable effects and the lack of drugs with specific antagonistic properties for amino acid excitation may make this suggestion somewhat difficult to pursue. Recent reports (MAO et al., 1974a, b, c) have indicated that drug treatments in vivo which alter amino acids in the cerebellar cortex also elevate cyclic GMP levels, and that this effect is blocked by benzodiazepines. These drug induced changes offer tantalizing suggestions that some amino acids could transmit interneuronally and also regulate cyclic nucleotide dynamics. However, in such experiments the intervals between drug injections and chemical sampling are orders of magnitude longer than synaptic transmission times, and several intervening neuronal groups could underlie the effects measured in the cerebellum. In addition to a possible effect as an antagonist of the amino acid receptors which activate accumulation of cyclic GMP, benzodiazepines are also included in the growing list of drugs capable of modifying phosphodiesterase activity (DALTON et al., 1974; see WEINRUB et al., 1972).

Despite the fact that brain contains many polypeptides (see review, BLOOM, 1972) and that some hypothalamic releasing factors can alter the cyclic nucleotide dynamics of the pituitary (BURGUS and GUILLEMIN, 1970; BORGEAT et al., 1972; BRAZEAU et al., 1973), there have been no reports thus far indicating that natural peptides can increase either cyclic AMP or cyclic GMP in brain slices or homogenates.

C. Electrophysiological Assessment of Cyclic Nucleotide-Mediated Events

In applying the "second messenger" conceptualization of the role of cyclic AMP (SUTHERLAND et al., 1965), norepinephrine or dopamine would presumably act at certain surface receptors to activate the synthesis of cyclic AMP within the postsynaptic neurons. The intracellular cyclic AMP would then activate subsequent enzymatic or molecular events which, among other actions, could result in the changes in cell discharge rate observed when the norepinephrine or dopamine are applied iontophoretically or released by natural synaptic inputs.

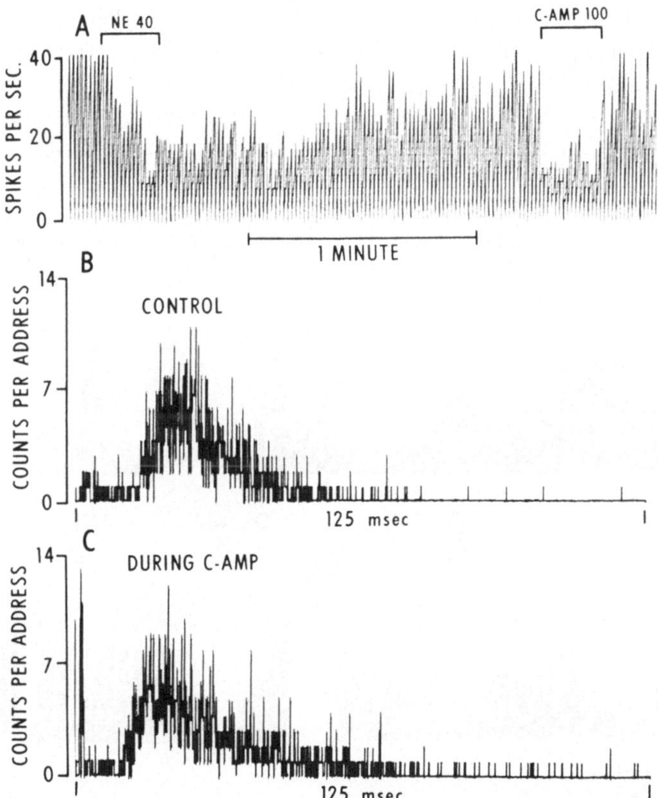

Fig. 3A–C. Slowing of spontaneous Purkinje discharges by iontophoresis of 40 nA of norepinephrine and 100 nA of cyclic AMP (C-AMP). A Polygraph record of mean discharge rate with time. Note more rapid onset and termination of depressant response with cyclic AMP than with NE. In this and all succeeding polygraph figures, the vertical pen deflection is proportional to discharge rate and the line accompanying microiontophoretic administration of drugs indicate the duration of drug application. The associated numbers represent the iontophoretic currents in nano-amperes. B and C Interspike interval histograms of spontaneous activity during control periods and during several submaximal applications of cyclic AMP, respectively. The peak of the histograms demonstrate the modal (most probable) intervals to be nearly identical (about 16 msec) in both B and C. In each histogram 1000 spikes were deposited in 500 addresses, each of 0.25 msec interval. (From SIGGINS et al., 1971a; with permission of Elsevier and Brain Research)

The innovative technique of microiontophoresis (see CURTIS, 1964; SALMOIRAGHI and BLOOM, 1964; BLOOM, 1974) has been quite useful in the assessment and identification of central synaptic transmitter substances, because drugs and other materials with charged groups can be applied from multi-barreled micropipettes in minute quantities directly adjacent to single neurons. Iontophoretic administration thus eliminates many of the diffusional and enzymatic barriers which restrict the access of drugs to neuronal receptors when the administration route is parenteral or topical.

When cyclic AMP was first proposed as the mediator of the iontophoretic responses to norepinephrine of cerebellar Purkinje cells (SIGGINS et al., 1969; Fig. 3, 4), emphasis was placed on the actions of iontophoretically applied exogenous cyclic AMP. In the majority of cells tested, but not all, iontophoretic administration of cyclic AMP could produce changes in cerebellar Purkinje-cell discharge rate and pattern identical to the effects of norepinephrine (HOFFER et al., 1971b, 1972, 1973; SIGGINS et al., 1971a). Cyclic AMP also inhibits hippocampal pyramidal cells (SEGAL and BLOOM, 1974a; Fig. 4) cerebral pyramidal tract cells (PHILLIS et al., 1975; STONE et al., 1975) and caudate nucleus neurons (SIGGINS et al., 1974).

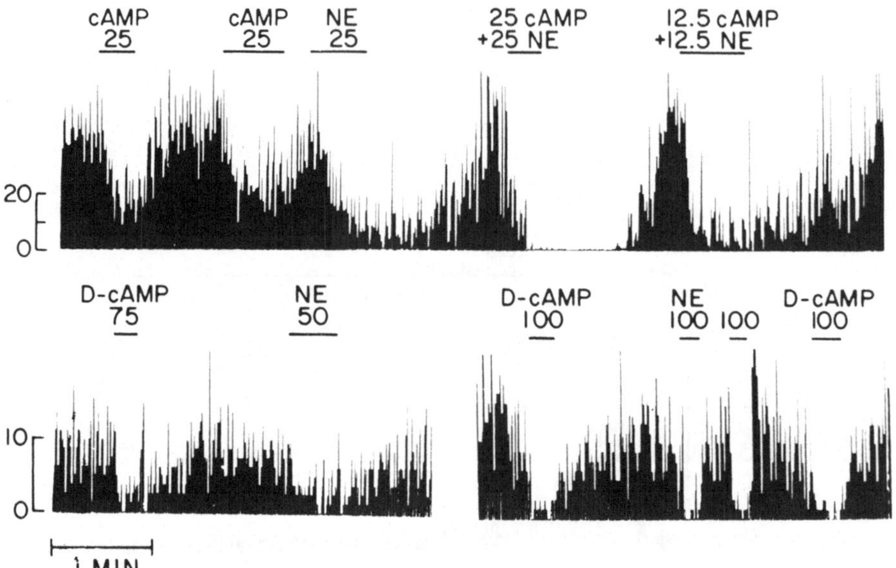

Fig. 4. Responses of spontaneously active hippocampal pyramidal cells to iontophoretic application of C-AMP, dibutyryl C-AMP (D-cAMP) and norepinephrine (NE). In this and other records, the neuronal firing rate is indicated by a polygraph in which the vertical height of the pen deflection is proportional to the number of spikes recorded in one second. Drug application periods are indicated by a horizontal bar above the polygraph record, with the magnitude of the ejection current expressed in nanoA. Three different test cells are illustrated. Above: C-AMP alone causes less prolonged and less intense inhibition of discharge than an equivalent ejection current passed through the NE barrel; when NE and C-AMP ejections are combined, a potentiative interaction is seen on both the time and intensity of the response which exceeds control periods even when both ejection currents are reduced by half. Below, left and right: Roughly equivalent responses to dibutyryl C-AMP and to NE. (From SEGAL and BLOOM, 1974a, with permission of Brain Research)

However, in order to interpret the results of experiments in which iontophoresis is used to evaluate a given compound, several technical and biological drawbacks of the system should be understood before proceeding to a consideration of the effects of cyclic nucleotides on neuronal firing rates and membrane properties.

1. Technical Problems in the Interpretation of Iontophoretic Tests of Cyclic Nucleotides

In order to assess the relative potencies of cyclic nucleotides and the transmitter substances whose synaptic actions the nucleotides might mediate, it is necessary to determine a characteristic property of compounds delivered by microiontophoresis, namely the mathematical constant, or transfer number, which relates the amount of substance released from the micropipet and the magnitude and duration of the iontophoretic delivery current. By studying the in vitro release of radiolabeled compounds of high specific activity (OBATA et al., 1970; HILL and SIMMONDS, 1973; BRADSHAW et al., 1973) and using brain pieces as a collecting medium (see SHOEMAKER et al., 1975), the transfer number for cyclic AMP has

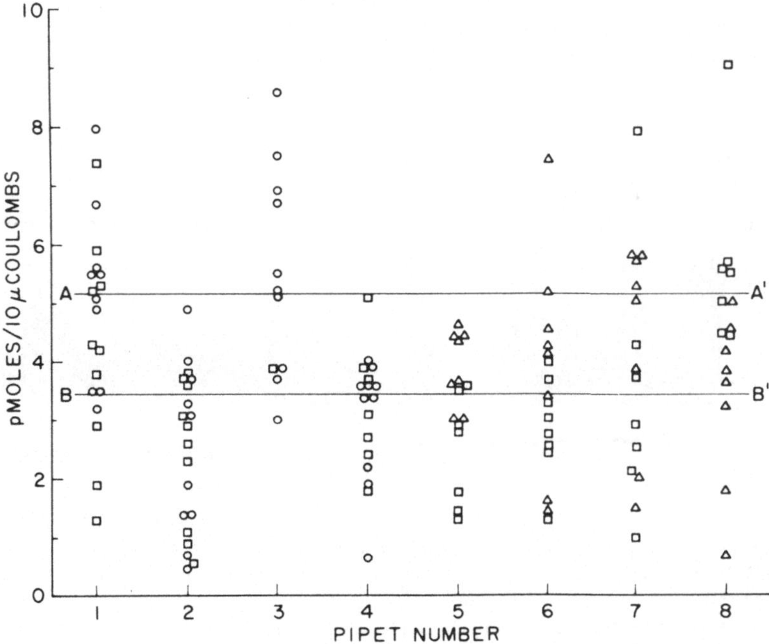

Fig. 5. Normalization of all data from eight pipets plotted as pmoles of cAMP released per 10 μcoulombes of current (e.g., 8 pmoles released by 20 μc is graphed as 4 pmoles). Squares (□) and triangles (△) represent samples obtained from two different iontophoresis circuits both of which maintain constant current by means of vacuum-tube photocells using 190 volt batteries. Circles (○) represent samples obtained on a third iontophoresis apparatus which utilizes operational amplifiers wired as current pumps powered by line current through a D.C. converter. 20% of all points lie above line A-A', 60% of all points lie above line B-B'. (From SHOEMAKER et al., 1975, with permission of Raven Press)

been determined with ejection currents and delivery periods approximating those used during in vivo experiments.

Under these in vitro conditions, the release of cyclic AMP was linearly related to both iontophoretic current intensity and time. The results revealed that cyclic AMP has a rather low transfer constant (0.05) relative to synaptic transmitter substances (generally, 0.10 or more). In addition, an unusually large amount of variation of release, both within and among pipets was found under a variety of times and currents (Fig. 5). The cause of the variation is not yet known but could be due to the unusual structure of the cyclic molecule and the fact that it must be iontophoresed as a negative ion. The large variation in iontophoretic release of cyclic AMP between trials of the same pipet and among different pipets stands in sharp contrast to the more regular release of other active molecules. Therefore, if a particular in vivo experimental situation required more than 5 pmoles of cyclic AMP to elicit a physiological response, for example, only about 20% of the iontophoretic trials would deliver an adequate dose (those points lying above line A–A′), and many pipets would never give a positive response. Similarly, and consequently, a physiological response requiring only 3.5 pmoles might only be seen 60% of the time (those points which lie above line B–B′) if adequate drug delivery were the exclusive limiting factor.

However, several other technical factors also enter into the interpretation of iontophoretic tests, including such parameters as the reliability of constant iontophoretic currents, the range of current values explored in attempts to reach

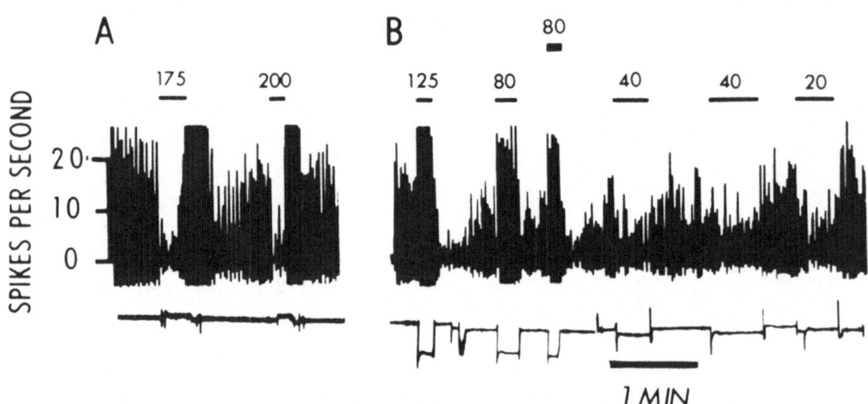

Fig. 6. Responses of a spontaneously active cerebellar Purkinje cell to iontophoretic administration of cyclic AMP and synthetic derivatives. Two types of data are illustrated: polygraphic recording of discharge rate, and (below) the continuous display of the magnitude of neutralization current "imbalance" at the pipet tip during and between tests with cyclic AMP. In A, monobutyryl C-AMP produces profound inhibition of spontaneous discharge at 175 and 200 nanoA of ejection current and virtually no neutralization "imbalance" during the trials. Note the marked excitatory rebound following the termination of monobutyryl C-AMP (thin horizontal lines). In B, on same cell as A, dibutyryl cyclic AMP (thick horizontal line) and high currents of monobutyryl cyclic AMP (thin horizontal lines) give complex responses. Note that attempts to eject C-AMP or dibutyryl C-AMP at 80–125 nA are accompanied by negative current imbalances (50–60 nA) and elicit apparent excitatory responses. As the ejection current of monobutyryl C-AMP was decreased to 40 or 20 nanoA, imbalance was reduced to 10 nA or less and the underlying C-AMP-induced inhibitory responses could then be seen. The best inhibition is seen at a level of C-AMP ejection (20 nA) which shows no imbalance. (G.R. SIGGINS, unpublished record; from BLOOM et al., 1975 with permission of Raven Press)

threshold effects, and the exclusion from such tests of the artefactual results of the iontophoretic current per se (see Fig. 6). These current effects may arise from tests in which the algebraic sum of the currents being used to eject one substance does not equal those being used to retain the other drugs within the multibarrel pipet. This imbalance leads to a local potential field at the electrode tip which can directly influence neuronal activity. While iontophoretic current control devices do exist which can automatically minimize the net current flowing between the multi-barrel electrode tip and ground, such devices are not routinely employed (see BLOOM, 1974). The complete elimination of ejection current imbalance at the electrode tip is essential, since the anionic currents used to deliver cyclic AMP, will, per se, often excite cells directly (see Fig. 6; also HOFFER et al., 1971a; FREEDMAN and HOFFER, 1974).

Therefore, when comparing iontophoretic data of responses to cyclic nucleotides, results from different laboratories should be evaluated in terms of the stated criteria by which responses were pursued and evaluated. For example, our definition (see SIGGINS et al., 1969; HOFFER et al., 1971a, b; SIGGINS et al., 1971a, b, c, d; HOFFER et al., 1972, 1973; BLOOM et al., 1974a; BLOOM, 1974) of non-responsive neurons required that at least two tests could be accomplished in which at least 200 nA of ejection current were applied without significant net tip current imbalance.

2. Biologic Problems in the Interpretation of Cyclic Nucleotide Tests on Neuronal Activity

In addition to technical problems associated with the variable iontophoretic release of cyclic AMP from pipets, certain biologic considerations in such tests are also important. Thus, the status of the animal or surgical exposure of the brain may result in transient anoxia or hypercapnia, or to hemorrhages or edema on the cortical surface. As a result, the neuronal discharges recorded may be abnormal or absent without exogenous excitants. Such activity may be very difficult to influence with cyclic AMP, especially with rapidly discharging cells excited by amino acids. Thus, depressant actions of cyclic AMP may be missed during the excitatory rebound which follows the application of almost all substances which depress rapidly firing cells (see LAKE and JORDAN, 1974; BLOOM et al., 1974a). These covert biological variations can impose an experimental ambiguity which results in confusing or non-reproducible data.

At least in concept, iontophoretic evaluation of the effects of exogenous cyclic nucleotides also presents a biologic problem with respect to the presumed site of their action. Under the second messenger hypothesis these substances would presumably act within the cell, perhaps through a protein kinase. Although extracellular iontophoresis of first messenger neurotransmitter candidates can successfully probe external receptivity, the method may be inadequate to probe intracellular actions. For example, in cell types in which threshold doses for first and second messengers have been compared (see ROBISON et al., 1971), the threshold concentration of cyclic AMP usually exceeds that of the hormone by 100 to 10000 fold.

One final biologic point concerns the ability to identify the appropriate target cell for the iontophoretic tests of cyclic nucleotides. One of the major attributes of the electrophysiological approach is the ability to examine the effects of a test substance on a single neuron or particular type of synaptic potential, for example: the action of a cyclic nucleotide may be compared with the effects of a particular neurotransmitter or nerve pathway which the cyclic nucleotide might be suspected to mediate. This approach takes on added importance when we recall that one of the interpretive shortcomings of studying the dynamics of cyclic nucleotides in brain slices is the inability to discern the responsive cell type when those responses might be diluted by a far greater number of unresponsive cells. Furthermore, by demonstrating the cytological and physiological identity of the cells which have been tested, data on homogeneous populations of test cells can be accumulated, a feature particularly useful when multiple pharmacological tests are to be compared.

3. Actions of Cyclic Nucleotides on Neuronal Firing

Having pointed out some of the major conditions under which the data obtained by microiontophoretic tests of cyclic nucleotides must be interpreted, the results reported by this method for cyclic AMP (Table 4) and cyclic GMP (Table 3) can be summarized rather quickly.

Cyclic AMP has been tested on identified cell populations in the cerebellum, hippocampus and cerebral cortex, and on other populations of target cells known to receive a high density of dopamine-containing nerve terminals in the caudate nucleus and the limbic forebrain (see Table 4). In each of these cases, in which the effects of the natural catecholamine are known to be inhibitory (see BLOOM and HOFFER, 1973), the major response to cyclic AMP is depression of from 20–70% of the cells tested. The low proportion results have in general stemmed from studies in which the criteria for cell identification or for appropriateness of testing were less rigorously controlled. In some tests, cells were not identified at all and current imbalance was not actively regulated. Since some definite disagreement exists as to exactly what proper criteria for such tests may be, reference is made to reviews of opposing view points (BLOOM, 1974; PHILLIS, 1974a, b; LAKE and JORDAN, 1974; BLOOM et al., 1974a).

Despite these controversial approaches, it now appears that a much higher proportion of spontaneously active, identified neurons in cerebral cortex are depressed by cyclic AMP (PHILLIS et al., 1974, 1975; STONE et al., 1975) than were unidentified cells activated by glutamate (see LAKE et al., 1972, 1973; PHILLIS, 1974a).

Cyclic GMP has been tested in some detail by iontophoresis on only two populations of brain cells (Table 3), the pyramidal tract neurons of the rat cerebral cortex (STONE et al., 1975) and a heterogeneous population of unidentified cat cerebro-cortical neurons (PHILLIS, 1974b). In the former case of identified cells, the responses to cyclic GMP bear a good correlation to qualitatively similar cholinergic muscarinic responses in being mainly excitatory (see STONE, 1974), while in the latter case, the effects of cyclic GMP bore no relation to the effects of acetylcholine on the same unidentified cortical cells, although most of these cells responsive to cyclic GMP showed excitation. In other tests (SIGGINS et al., 1971a; HOFFER et al., 1971b; PHILLIS et al., 1975) cyclic GMP has been reported as mainly a depressant.

Table 3. Effects of cyclic GMP administered by Iontophoresis to central neurons

Cell Type	N	% Responding			Reference
		R	S	O	
Rat Cerebral Cortex					
Pyramidal tract	41	21	49	30	STONE et al., 1975
Cat Cerebral Cortex					
Unidentified	32	3	51	42	PHILLIS, 1974

Key: N = Number of cells tested; R = Percentage of tested cells reduced in rate; S = Percentage of tested cells speeded; O = Percentage of tested cells with no response.

Table 4. Effects of cyclic AMP administered by iontophoresis to central neurons

Cell Type	N	% Responding				References
		R	S	O	V	
Cerebellum						
Rat Purkinje Cells-Normal	137	64	11	17	8	HOFFER et al., 1971a, b, 1972, 1973; SIGGINS et al., 1971a–d
Rat Purkinje Cells-Normal	75	8	34	57	0	GODFRAIND and PUMAIN, 1971
Rat Purkinje Cells-Normal	31	19	32	49	0	LAKE and JORDAN, 1974
Rat Purkinje Cells-Irradiated	42	74	7	19	0	WOODWARD et al., 1974
Rat Purkinje Cells-6-OHDA	19	58	16	26	0	HOFFER et al., 1971c
Rat Purkinje Cells-Neonatal	7	100	–	–	0	WOODWARD et al., 1971
Cat Purkinje Cells	20	65	5	30	0	HOFFER, SIGGINS, BLOOM, unpublished
Frog Purkinje Cells	5	60	–	40	0	HOFFER, SIGGINS, BLOOM, unpublished
Pigeon Purkinje Cells	8	50	12	25	12	HOFFER, SIGGINS, BLOOM, unpublished
Hippocampus						
Rat Pyramidal Cells	44	57	2	41	0	SEGAL and BLOOM, 1974a, b
Caudate Nucleus						
Rat (Unidentified)	88	86	3	11	0	SIGGINS et al., 1974; BUNNEY and AGHAJANIAN, 1974
Cerebral Cortex						
Rat Pyramidal Tract	18	66	22	11	0	STONE et al., 1975
Rat (Unidentified)	18	11	6	83	0	LAKE et al., 1973
Cat (Unidentified)	43	16	9	75	0	LAKE et al., 1973
Guinea Pig (Unidentified)	23	4	13	83	0	LAKE et al., 1973
Limbic System						
Rat (Unidentified)	27	89	0	11	0	BUNNEY and AGHAJANIAN, 1964
Brain Stem						
Cat (Unidentified)	68	79	0	21	0	ANDERSON E.G. et al., 1973

Key: N = Number of cells tested; R = Percentage of tested cells reduced in rate; S = Percentage of tested cells speeded; O = Percentage of tested cells with no response; V = Percentage of tested cells with variable or biphasic responses (SIGGINS et al., 1969).

4. Effects of Cyclic Nucleotides on Membrane Properties of Neurons

At the present time, the most detailed comparison which can be made between the effects of a cyclic nucleotide and the effects of a neurotransmitter or a nerve pathway is with respect to the membrane properties of the test cell. Through the use of an intracellular electrode with intracellular or extracellular iontophoretic drug application (Fig. 7), or through the use of the sucrose gap method of membrane potential recording (see WEIGHT, 1974), it is possible to compare the magnitude and direction of the change in membrane potential accompanying the response to the applied substance or to stimulation of the nerve pathway (Fig. 8). Both methods can permit some insight into the ionic conductance changes which occur during these responses. It should be clear, however, that the same technical precautions of extracellular iontophoretic studies apply here as do additional technical requirements imposed by the even more difficult evaluation of transmembrane properties.

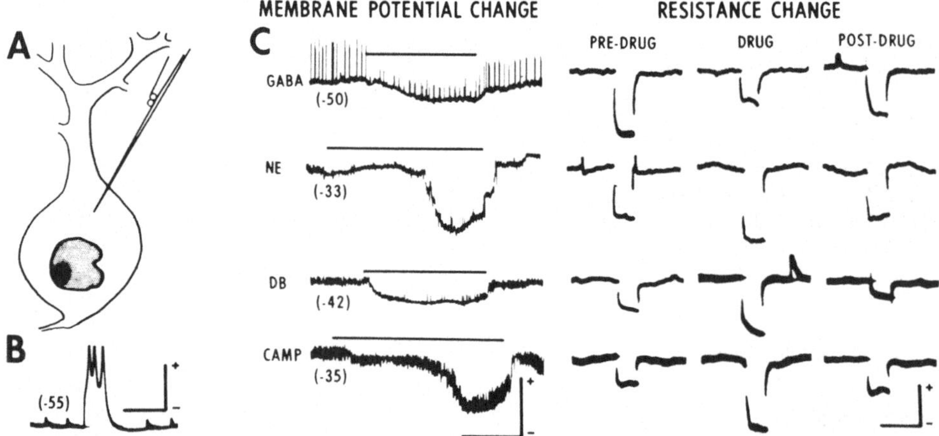

Fig. 7A–C. Intracellular recordings from rat cerebellar Purkinje cells. A Schematic representation of a three-barrel micropipet with a Purkinje cell. The intracellular electrode protrudes beyond the orifices of the two extracellular microelectrophoresis barrels. B Multispiked spontaneous climbing fiber-evoked discharge obtained during intracellular recording from a Purkinje cell. Number in parentheses is resting potential in millivolts (mv); calibration bars are 20 msec and 25 mv. C Changes in membrane potential and membrane resistance of four different Purkinje cells in response to gamma aminobutyrate (GABA), norepinephrine (NE), dibutyryl cyclic AMP (DB), and cyclic AMP. All specimens in each horizontal row of records are from the same cell. Solid bar above each record indicates the extracellular electrophoresis of the indicated drug (100–150 nA). Number in parentheses below each recording is resting potential in millivolts; calibration bar under membrane potential records is 10 seconds and 20 mv for NE, DB cyclic AMP, and cyclic AMP, and is 5 seconds and 10 mv for GABA. The effective input resistance was judged by the size of pulses resulting from the passage across the membrane of a brief constant current (1 nA) pulse before, during, and after electrophoresis of the respective drugs (1 mv = 1 megohm). Discontinuities in the fast transients of the pulses result from the loss of high frequencies (> 10 kHz) and from the chopped nature of the frequency-modulated magnetic tape recording used. All "pulse" records were graphically normalized to the same baseline level. Calibration bar on right indicates 80 msec and 15 mv for all pulse records. (From SIGGINS et al., 1971b with permission of the American Association for the Advancement of Science)

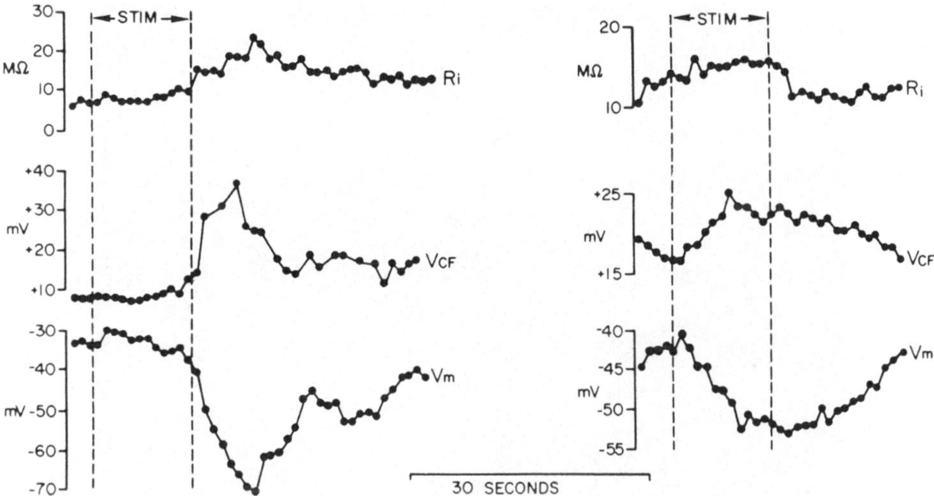

Fig. 8. Intracellular responses of a Purkinje cell to stimulation of locus coeruleus with multiple shocks (STIM). Both panels represent digital computer display of records taken from the same cell. Left panel recorded immediately after penetration of the cell. Right panel was recorded five minutes later. In each case the stimulus was 120 pulses of 0.45 mA (supramaximal intensity) delivered at 10/sec. Records were analyzed at one-second intervals by a PDP-12 computer and the input resistance (Ri), magnitude of climbing fiber-evoked responses (V_{CF}) and membrane potential (Vm) plotted against time. Note the change in vertical calibration from the left to right panel and the variability in magnitude typical of locus coeruleus responses. (From SIGGINS et al., 1971c, with permission of Nature)

D. Other Methods for the Assessment of Cyclic Nucleotide-Mediated Events

A third methodological approach toward defining the functional role of cyclic nucleotides in the nervous system derives from histochemistry. In this approach, fresh, frozen, or partially fixed sections of nervous tissues are examined by light or electron microscopy. Chemical or immunochemical reactions are performed on the tissue to determine where a particular substrate or enzyme activity is located. Similarly, these methods can demonstrate whether the substrate or enzyme reactivity can be altered by experimental procedures (see below). While this approach can clearly impart some useful information which is presently impossible to resolve with other techniques, the cost of this information is the requirement for a wide range of control reactions to establish specificity and selectivity of the reaction. In some cases, this price has not yet been fully paid.

1. Immunocytochemical Localization of Cyclic Nucleotides

When cyclic AMP or cyclic GMP are coupled to large carrier proteins, the resultant haptene molecule can be employed as an antigen to raise antibodies which will selectively react with either cyclic AMP or cyclic GMP specifically

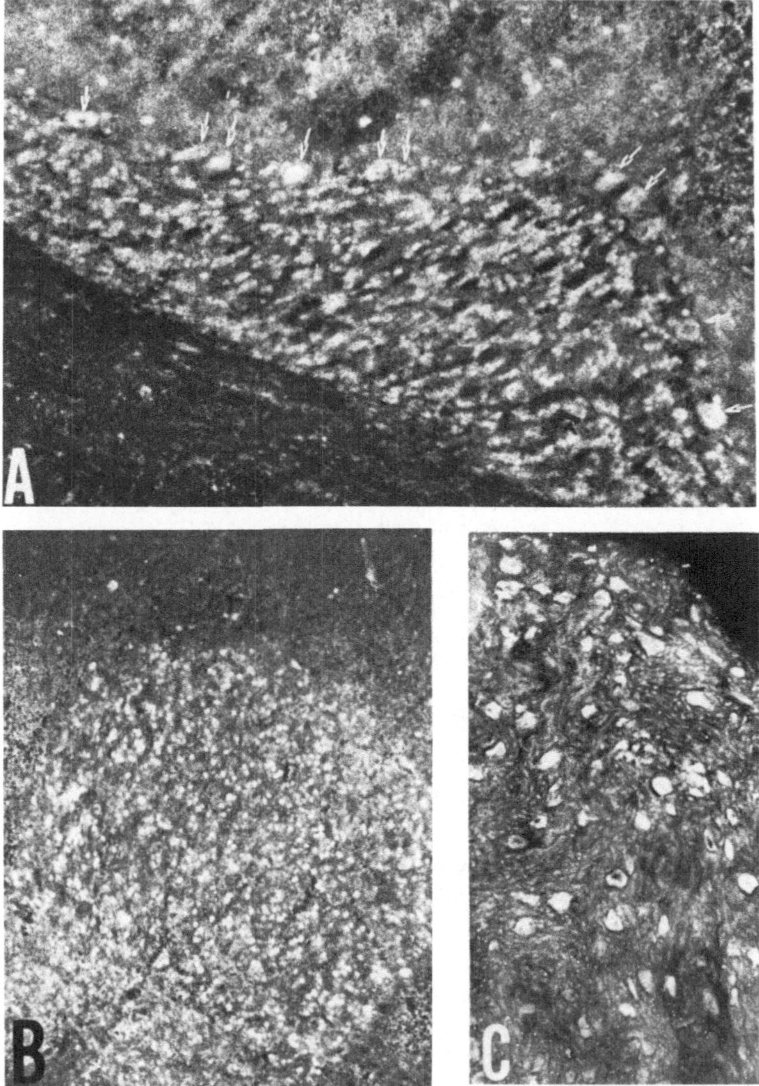

Fig. 9A–C. Dark-field fluorescence micrographs showing the immunocytochemical staining patterns for intracellular cyclic AMP in rat cerebellum (A and B) and in bovine sympathetic ganglion (C). A The staining pattern in a frozen biopsy specimen taken approximately 150 sec after decapitation comparable to the peak rise in cyclic AMP tissue levels. Low reactivity is seen in the white matter (lower left) or molecular layer (upper right) but perikarya of Purkinje cells (arrows) and the nuclei of granule cells (between Purkinje cell layer and white matter) show strong reactivity. B Is similar to A except that the specimen was frozen within 30 sec of decapitation, before cyclic AMP levels are significantly elevated; note that only granule cell nuclei show reactivity and that Purkinje perikarya are unstained. C The intense staining of the large post-ganglionic neurons of the bovine sympathetic ganglion after an exposure to 100 μM dopamine in vitro; neuronal cytoplasm is intensely reactive, while intercellular material shows little reactivity. (A and B from BLOOM et al., 1972, unpublished; C from KEBABIAN, BLOOM, STEINER and GREENGARD, unpublished)

(see STEINER et al., 1970, 1972a, b). In addition to the original application in the radio-immunoassay of cyclic AMP and cyclic GMP (STEINER et al., 1970, 1972a–c), the immunoglobulin fractions of the anti-cyclic AMP antisera can be employed for purposes of immunocytochemistry to localize cyclic AMP (WEDNER et al., 1972) or cyclic GMP (FALLON et al., 1974) bound within cells. This technique has been employed to examine the location of cyclic AMP within cells of the cerebellar cortex under resting conditions and under experimental conditions which are known to induce an elevation in the cerebellar content of cyclic AMP. When the phenomenon of post-decapitation anoxia (BRECKENRIDGE, 1964) was used to elevate cerebellar cyclic AMP levels, immunofluorescent staining for cyclic AMP revealed only two types of reactive cells: granule cells and cerebellar Purkinje cells (BLOOM et al., 1972; Fig. 9A). When cerebellar samples were examined after decapitation, but before the onset of the major rise in cyclic AMP concentration, only granule cells showed immuno-reactivity (BLOOM et al., 1972; Fig. 9B). All general anesthetics except halothane also abolished the staining increase in Purkinje cells as they do the post-decapitation rise in levels (see BLOOM et al., 1972; KIMURA et al., 1974).

In order to determine the staining patterns in cerebellum under conditions which could more closely approximate the electrophysiological studies, cerebellar biopsies were taken from anesthetized rats prepared as for experiments in electrophysiology. The exposed cerebellar cortex has been used to test the actions of topically applied neurohormones and of the norepinephrine synaptic pathway to the cerebellum as revealed by changes in the immunoreactivity to the anti-cyclic AMP immunoglobulin (SIGGINS et al., 1973). Topical application of NE in concentrations of 10–100 μM or electrical stimulation of locus coeruleus resulted in nearly a 5–7 fold increase in the number of immunoreactive cerebellar Purkinje cells (from 10% to more than 70% reactivity) (Fig. 10); much higher concentrations of other inhibitory substances did not increase the immunoreactivity of cerebellar Purkinje cells to anti-cyclic AMP immunoglobulin. The increases in immunoreactivity of cerebellar Purkinje cells to locus coeruleus stimulation could be blocked if the locus coeruleus-cerebellar pathway was first destroyed by treatment with 6-hydroxydopamine (see SIGGINS et al., 1973). The immunocytochemical approach thus was able to provide some direct indication that the cyclic AMP content of cerebellar Purkinje cells can be increased in response to applied norepinephrine or to activation of the norepinephrine-containing locus coeruleus pathway.

The immunocytochemical approach to the localization of bound cyclic nucleotides has also been applied to the bovine sympathetic ganglion preparation (see GREENGARD and KEBABIAN, 1974). After exposure to dopamine, cyclic AMP immunoreactivity increases almost exclusively in the ganglion neurons while immunoreactivity to cyclic GMP rises only after exposure to acetylcholine (KEBABIAN, BLOOM, STEINER and GREENGARD, unpublished observations; Fig. 9C).

The specificity of this immunoreaction has been partially established through such experiments as those described above which indicate that reactivity for a given nucleotide is generally restricted to particular cell types within a sample of nervous tissue, and that the degree of immunoreactivity increases under conditions known to result in increased tissue cyclic nucleotide levels. However, the

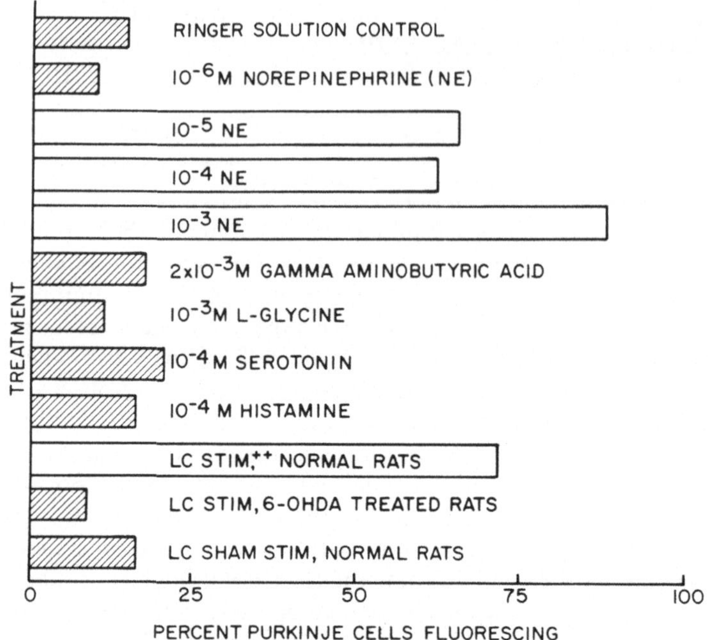

Fig. 10. Effects of pharmacological and electrophysiological stimuli on the immunoreactivity of cerebellar Purkinje cells for cyclic AMP. Solutions were applied by cotton pledget to the exposed cerebellar vermis for 10 min and then frozen biopsies were taken for immunocytochemical staining. Results are expressed as the mean percent of reactive Purkinje cells, which was determined by dividing the number of reactive cells by the total number of Purkinje cells in individual folia. The percentages of each sampled folia were averaged for groups of three to eight folia in at least two separate experiments to provide standard deviations. More than 200 cells were counted for each condition. Hatched bars indicate values that are not significantly different from Ringer solution controls ($P > 0.05$; Student's t-test); clear bars indicate values significantly greater than control values ($P \wedge 0.001$); LC Stim, electrical stimulation of ipsilateral locus coeruleus; 6-OHDA, 6-hydroxydopamine. (From SIGGINS et al., 1973; with permission of the American Association for the Advancement of Science)

specificity for the reaction has also been further established through the use of control incubations on salivary gland, pineal, toad bladder and leucocytes, as well as peripheral and central neurons. Here one or another of the immunocomponents is either omitted or pre-reacted with blocking concentrations of the cyclic nucleotide antigen (see WEDNER et al., 1972; FALLON et al., 1974). The results of these tests indicate that bound intracellular cyclic nucleotides in properly prepared tissues can resist extensive washing and therefore can be localized, presumably by virtue of their binding to the regulatory protein of the cyclic nucleotide-dependent protein kinase. However, there has not yet been a comprehensive demonstration of the amount of cyclic nucleotide eluted from the frozen, dried tissue section during the processing, nor has the method yet been made quantitative. Nevertheless, qualitative estimations of the location of cyclic nucleotides in the presence of a transmitter which activates the synthesis of a given cyclic nucleotide does seem to provide useful information as to which cells in a region are the most responsive cell types.

2. Cytochemical Localization of Adenylate Cyclase

A general method for the electron microscopic localization of hormonally-responsive adenylate cyclase activity was first described by REIK et al. (1970). Their method employed ATP as a substrate, trapping the pyrophosphate liberated from the cyclase reaction with Pb^{++}; the Pb^{++} precipitate was subsequently resolved as an electron dense particle along the surface of rat hepatocyte membranes. Although ATP can also be catabolized by many hepatic enzymes to liberate either phosphate or pyrophosphate, the fact that the enzyme reactivity observed by REIK et al. (1970) was increased greatly upon exposure of the tissue to glucagon, to isoproterenol, or to F^-, provided strong suggestive evidence that the reaction product was associated with adenylate cyclase.

By employing a substrate which is less likely to be catabolized by the more general tissue ATPase enzymes, namely adenylylimidodiphosphate (AMP-PNP) (see YOUNT et al., 1971) HOWELL and WHITFIELD (1972) demonstrated glucagon-activated adenylate cyclase reactivity in partially fixed, isolated, pancreatic islets. Almost simultaneously, WAGNER et al. (1972) used a similar procedure to demonstrate a catecholamine-stimulated adenylate cyclase reactivity in unfixed isolated capillary endothelial cells. In reviewing the protocols and interpretation of these procedures, WAGNER and BITENSKY (1974) also described the only reported attempt to localize adenylate cyclase in a neural tissue: the isolated rod outer segments of the frog retina. Their micrographs indicate that a reaction product, formed upon exposure to AMP-PNP and Pb^{++}, accumulated in the rod outer segments and that it was inhibited by alloxan. However, no increase in reactivity upon physiological or pharmacological stimulation was reported.

To establish that these cytochemical reactions demonstrated adenylate cyclase reactivity, most of the group have shown that the enzyme is active in their tissue samples under conditions analogous to those used for cytochemistry, with the exception that the Pb^{++} salt used to trap the pyrophosphate or imidodiphosphate was omitted. When ATP is used as the substrate, Pb^{++} must be omitted from biochemical controls since in the presence of Pb, ATP will spontaneously break down to cyclic AMP (REIK et al., 1970) and probably other phosphates as well (see MOSES and ROSENTHAL, 1968). As a result of this effect of Pb^{++}, it has not yet been established that the millimolar concentrations of Pb^{++} used to detect the cytochemical reaction product do not themselves interfere with basal or hormonally-stimulated adenylate cyclase. Such a demonstration would significantly strengthen this promising method and perhaps accelerate efforts to achieve localizations within other parts of the nervous system. The results obtained with this method at present suggest that at least the adenylate cyclase associated with the plasma membrane (see SUTHERLAND and RALL, 1962) may be localizable. If it were possible to adapt this method for neurons, the important relationship to be probed by this cytochemical approach would be the topological arrangement of membrane enzyme activity specifically activated by different transmitter substances relative to the location of their synapses.

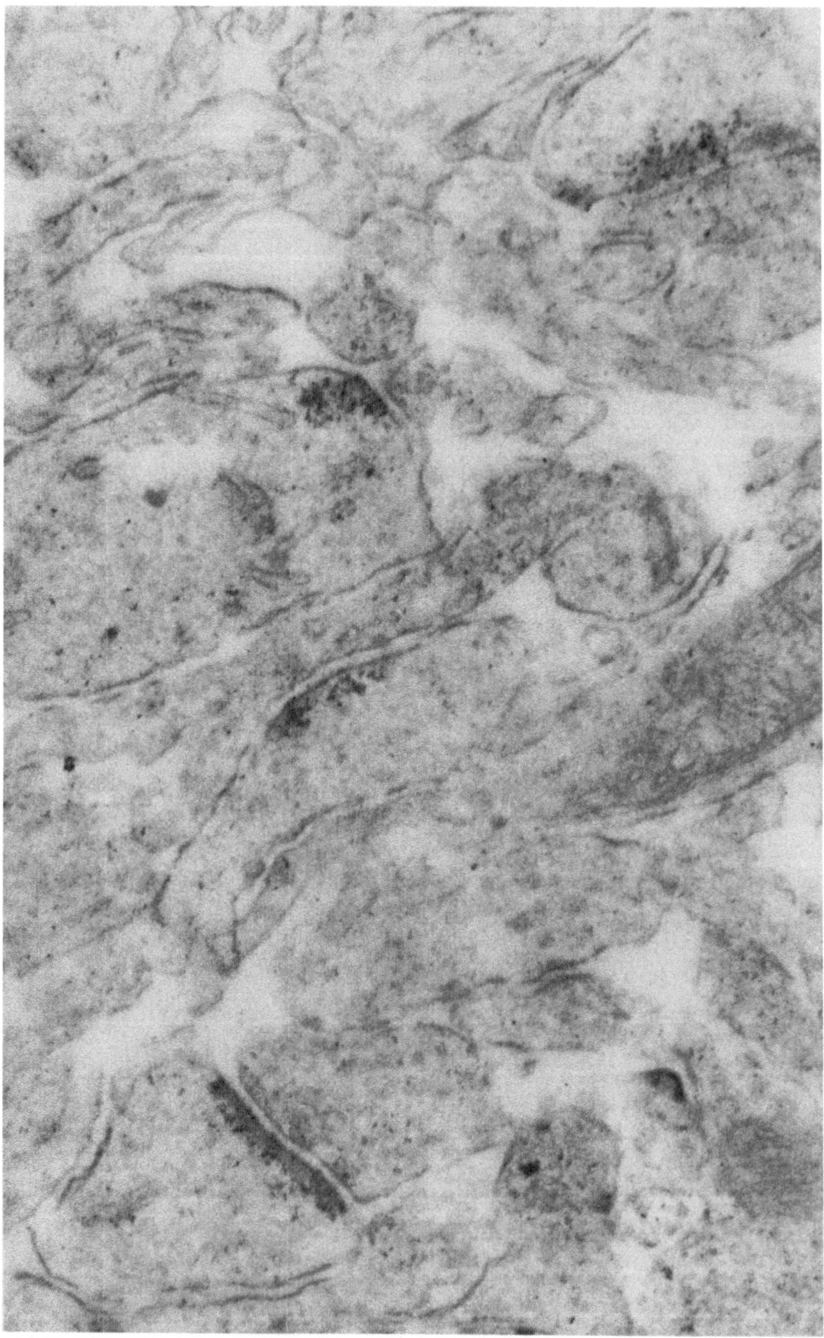

Fig. 11. Electron micrograph of mouse cerebral cortex (15 days postnatal development) reacted for localization of phosphodiesterase activity; electron-dense particulate deposits with the postsynaptic membrane of three small dendritic processes indicates that phosphodiesterase may be concentrated there (Mag. 84000; from ADINOLFI et al., 1974, with permission of the authors and Brain Research)

3. Cytochemical Localization of Phosphodiesterase

The localization of enzyme activities capable of hydrolyzing the 3′-bond of cyclic AMP has been approached with a reaction mixture similar to that used to localize adenylate cyclase, except that in this case cyclic AMP is added as the substrate. Upon reaction with the phosphodiesterase of the tissue sample, 5′-AMP is formed, which in turn reacts with a purified 5′-nucleotidase added as a second reagent. The inorganic phosphate liberated by this latter reaction is trapped by precipitation with Pb^{++} for visualization. Controls for such reactions consist of the omission of one or more of the reagents or the addition of an inhibitor of the phosphodiesterase, such as a methyl xanthine. Interestingly, the omission of the 5′-nucleotidase reagent resulted in no visible reaction product (FLORENDO et al., 1971); on the other hand, a phosphodiesterase inhibitor has been routinely added in the adenylate cyclase reaction mixture (WAGNER and BITENSKY, 1974) to prevent any endogenous 5′-nucleotidase reactivity from yielding an artefactual localization reaction precipitate from hydrolysis of 5′-AMP.

Using the cyclic AMP-Pb^{++}-5′-nucleotidase reaction procedure, FLORENDO et al. (1971) were the first to report that in the rat cerebral cortex, a theophylline-sensitive phosphodiesterase activity was localized primarily to the postsynaptic membrane specialization. More recently, ADINOLFI and SCHMIDT (1974) confirmed this general result and demonstrated a similar location of reaction product occurring during the stage of development in which cortical synapses are formed (Fig. 11). A postsynaptic localization of phosphodiesterase activity is in general confirmation of the quantitative histochemical measurements made in sequential frozen sections of several parts of the rat central and peripheral nervous system by BRECKENRIDGE and JOHNSTON (1969) and is compatible with the subcellular distribution studies of DE ROBERTIS et al. (1967). While the subsynaptic localization of phosphodiesterase activity could imply a cyclic nucleotide relationship to synaptic transmission, it has not yet been possible to test this inference by showing that this enzyme reactivity is present only in those types of synaptic connection for which there is physiological or pharmacological evidence of a cyclic nucleotide-mediated transmission. Furthermore, no quantitative comparisons between the frequency distribution of phosphodiesterase-positive synapses with regional phosphodiesterase activities has been achieved. If such localizations could be correlated, it might then be possible to attempt further functional correlations between the enzyme localized in this fashion and the different kinetic and pharmacological properties of the several phosphodiesterase isozymes known to be present in nervous tissue (see above, III. D.3).

V. Cyclic AMP and Catecholamine-Mediated Synaptic Events

The molecular analysis of function in the nervous system labors against a frustratingly complex system of heterogenous interconnected neurons and associated neuroglia. Because the relationships between these contiguous elements remains largely uncharted beyond the anatomical wiring diagrams of the major sensory and motor pathways, extensive analysis of the molecular basis of any given synaptic arrangement depends upon the extremely detailed anatomical and physiological specification of the presynaptic and postsynaptic cells under investigation. Therefore, the analysis of where and how cyclic nucleotides might participate in synaptic transmission must be undertaken along tactical lines in which the criteria for the identification of a synaptic transmitter (II.C) intersect with the criteria for establishing that the effects of a transmitter or hormone may be mediated by a cyclic nucleotide (III.A). From the data reviewed thus far, the central catecholamine-containing pathways would seem to represent the best available systems for this analysis since these pathways satisfy two practical requirements: (a) catecholamines are known to influence adenylate cyclase or cyclic AMP levels in various discrete regions of the nervous system by definable receptors and (b) the source neurons and target neurons of the central catecholamine pathways have been sufficiently characterized so that their effects can be determined and related to the effects of cyclic nucleotides and related substances. In this section, we examine in more depth the data derived from the intersection of the various technical approaches to the assessment of cyclic AMP by examining several specific catecholamine receptive target neurons.

A. Localization of Norepinephrine- and Dopamine-Containing Synapses

The major break-through in the understanding of where norepinephrine-containing neurons send their synaptic projections within the central nervous system has come from the development of a specific fluorescence histochemical method (see FUXE and JONSSON, 1973; UNGERSTEDT, 1971). Through progressive modifications and consistent applications, it is now possible to describe several discrete pathways in which the norepinephrine-containing neurons of the pontine nuclei, locus coeruleus and sub-coeruleus, extend axons to all cortical regions, including cerebral, cerebellar and limbic cortices (UNGERSTEDT, 1971; OLSON and FUXE, 1971; PICKEL et al., 1974a) and through different pathways to regions of the hypothalamus (OLSON and FUXE, 1972; also see LINDVALL and BJÖRKLUND, 1974).

However, the varicosities of axons revealed by the fluorescence histochemical method cannot adequately define the postsynaptic target cells in a given region, and electron microscopic methods, described in detail elsewhere (see BLOOM, 1973), are required for this purpose. Because no single method has so far proven capable of resolving all of the norepinephrine-containing terminals within a region, it is usually desirable to apply at least two of the following methods: (a) permanganate fixation, to produce small granular vesicles directly; (b) electron microscopy with autoradiography of terminals which accumulate H^3 norepinephrine or, (c) electron microscopy of terminals which are labeled after the microinjection into the area of the locus coeruleus of H^3 proline or leucine (these amino acids are presumably incorporated into proteins and carried to the terminals by axoplasmic transport; see PICKEL et al., 1974a); (d) location of terminals which are undergoing degeneration within 24–48 hours of the intraventricular or intracisternal injection of the selective neurotoxin, 6-hydroxydopamine (see BLOOM, 1971).

Synaptic terminals presumed to contain norepinephrine have been identified within the rat cerebellar cortex by a combination of fluorescence histochemistry, electron microscopic autoradiography after intracisternal H^3 norepinephrine injection, and the pattern of acute degeneration after intracisternal injection of 6-hydroxydopamine (BLOOM et al., 1971). Based upon the serial analysis of material prepared for formaldehyde-induced (OLSON and FUXE, 1971) or glyoxylic acid-induced fluorescence histochemistry (BLOOM and BATTENBERG, in preparation) and upon light microscopic autoradiography after microinjection of H^3 proline into the locus coeruleus (PICKEL et al., 1974a), the origin of these fibers is the nucleus locus coeruleus. The electron microscopic analysis of the cerebellum indicates the major target neurons to be the Purkinje cells (BLOOM et al., 1971). This identification is in part supported by recent observations on the mouse cerebellum in which permanganate fixation revealed small granular vesicles within terminals synapsing on the dendritic spines of Purkinje cells in normal and mutant mice (LANDIS and BLOOM, 1974; LANDIS, SHOEMAKER, SCHLUMPF and BLOOM, in preparation). On the other hand, PALAY and CHAN-PALAY (1974) have interpreted the rat data (BLOOM et al., 1971) differently, ascribing the terminals labelled by autoradiography of H^3 norepinephrine to parallel fibers, presumably because the labeled terminals did not exhibit the typical small granular vesicles seen in some central and almost all peripheral norepinephrine-containing terminals (see BLOOM, 1973). Small granular vesicles can be seen in boutons which contact Purkinje cell dendrites (LANDIS and BLOOM, 1974) with permanganate fixation, but this fixative is not compatible with autoradiography (see BLOOM, 1973). However, since it has been estimated that cytochemically defined terminals characterized as norepinephrine-containing by autoradiography or reaction to 6-hydroxydopamine may number less than 1 per thousand cerebellar cortical terminals (BLOOM et al., 1971), it is not surprising that simple analyses by routine descriptive methods of ultrastructural preparation have not yielded more significant numbers of these terminals. Further confirmation of the identification of the Purkinje cells as the synaptic targets of the locus coeruleus norepinephrine neurons (BLOOM et al., 1971) comes from the physiological and pharmacological studies described below (SIGGINS et al., 1971a–d, 1973; HOFFER et al., 1969, 1971, 1972, 1973; BLOOM et al., 1975).

Using similar light, fluorescence and electron microscopic methods, there is evidence that the pyramidal neurons of the hippocampus and dentate gyrus also receive norepinephrine containing synapses from the locus coeruleus (PICKEL, 1974a) and equally extensive evidence that the neurons of the rat caudate nucleus receive dopamine-containing synapses from neurons of the substantia nigra pars compacta (HÖKFELT and UNGERSTEDT, 1973; also see SIGGINS et al., 1974). There is also presumptive evidence that some neurons of the cerebral cortex of the squirrel monkey (NELSON et al., 1973) and of the rat (see FUXE, 1973; LINDVALL et al., 1974; BERGER et al., 1974) receive norepinephrine-containing afferents. In addition, a separate and distinctive projection of dopamine-containing synapses arising in neurons of the substantia nigra and the ventro-medial tegmentum (THIERRY et al., 1973; LINDVALL et al., 1974; FUXE et al., 1974; BERGER et al., 1974) and which extends to the rhinencephalic cortex, the temporal cortex and the anterior cingulate gyrus.

B. Norepinephrine-Mediated Synapses

1. Cerebellar Purkinje Cells and Hippocampal Pyramidal Cells as Norepinephrine Target Cells

Since the possible intraneuronal actions of cyclic AMP can now be assessed only in terms of acute changes in the biophysical properties of neurons, several experimental paradigms have been employed to gather indirect, but supportive evidence bearing on the question of whether cyclic AMP functions as a second intracellular mediator for norepinephrine synapses on cerebellar and hippocampal neurons. These paradigms include: (1) characterization of the catecholamine receptor and its similar function and possible synergism with cyclic AMP (HOFFER et al., 1971; SEGAL and BLOOM, 1974a; Fig. 13, 14); (2) the actions of phosphodiesterase inhibitors (SIGGINS et al., 1969, 1971a); (3) pharmacological interactions with agents such as the prostaglandins (see SIGGINS et al., 1971b) and certain phenothiazines (FREEDMAN et al., 1974) known to affect the responsiveness of brain adenylate cyclase (and of other tissues) to norepinephrine; (4) assessment of the effects of cyclic AMP on the membrane properties of the cerebellar Purkinje cell (SIGGINS et al., 1971c, d; HOFFER et al., 1973) and hippocampal pyramidal cells (OLIVER and SEGAL, 1974); and (5) the immunocytochemical localization of intracellular cyclic AMP (BLOOM et al., 1972; SIGGINS et al., 1973).

a) Characterization of the Receptors Activated by Norepinephrine and Stimulation of the Locus Coeruleus

In each of these test systems, norepinephrine produces inhibition of spontaneous discharge. Based upon the interactions with antagonists, the norepinephrine receptors on the cerebellar Purkinje cell and hippocampal pyramidal cells are characterized as beta adrenergic, since the responses to iontophoretic norepinephrine and to locus coeruleus stimulation are blocked by Sotalol or MJ-1999 (Fig. 12; see HOFFER et al., 1971; SEGAL and BLOOM, 1971a, b). The breadth of such studies

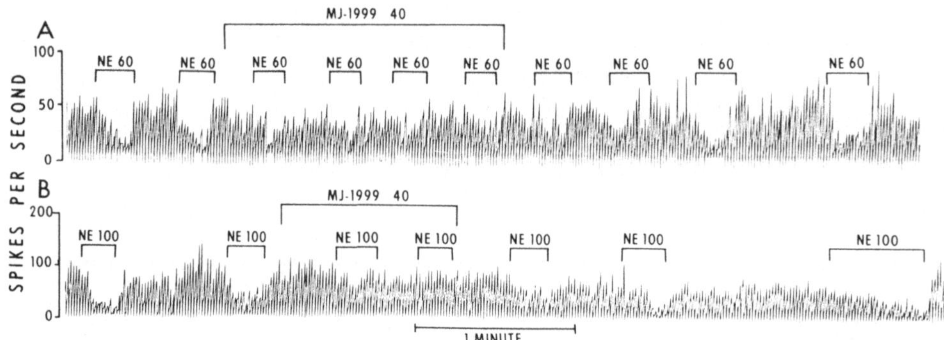

Fig. 12. Antagonism of Purkinje cell responses to NE by microiontophoresis of sotalol (MJ-1999). A and B represent 2 different cells. Note that MJ-1999 does not slow cells when applied alone, indicating the absence of local anesthetic action. NE was applied at lowest currents which produced maximal responses. (From HOFFER et al., 1971; with permission of Brain Research)

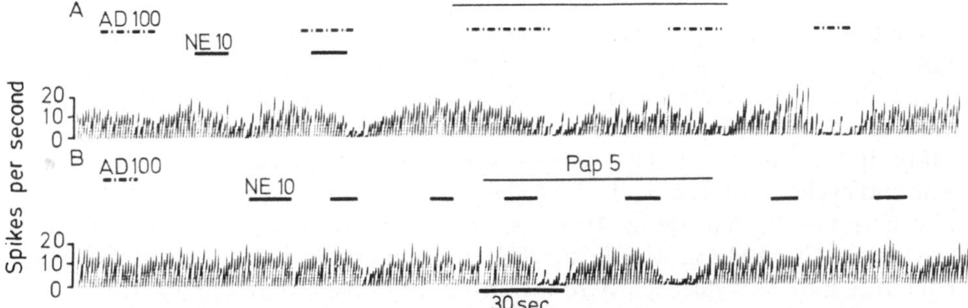

Fig. 13A and B. Interactions between NE, adenosine (AD) and papaverine on the spontaneous discharge of a rat cerebellar Purkinje cell. A When threshold ejection doses of adenosine and NE are passed simultaneously, the resultant response is only slightly better than to NE alone. However, the response to adenosine is greatly improved by the iontophoretic application of a level of papaverine which does not, per se, change basic discharge rate. B An even smaller ejection dose of papaverine markedly potentiates the response to NE given at regular repeated intervals. A and B are continuous records; but, note the apparent loss of responsiveness to NE during the ten min interval between the last NE test in A and the first NE test in B, due to continuous holding current uninterrupted by ejection currents. Papaverine has no overt actions on the discharge rate between NE trials. (G.R. SIGGINS, unpublished record; from BLOOM et al., 1975, with permission of Raven Press)

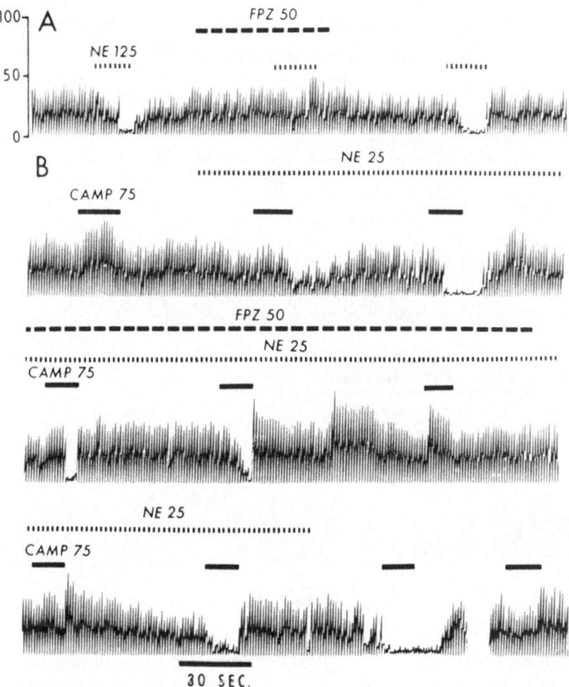

Fig. 14A and B. Discharge rates of a single Purkinje cell. A Rapid blockade of supramaximal NE (thin dotted line) by fluphenazine FPZ (thick dotted line). B Although Cyclic AMP (solid line) by itself has no inhibitory effect, subthreshold doses of NE and cyclic AMP interact to produce maximal inhibition, which is blocked by FPZ. Recovery of this interaction occurs after FPZ ejection is terminated. The three lines of B are continuous, except for a two min break in the last line. (From FREEDMAN and HOFFER, 1975, with permission of J. Neurobiology)

is limited since many potential antagonists, both alpha and beta types, can be directly depressant on test cell neuronal discharge. These studies provide the electrophysiological counterparts to the observations that norepinephrine is capable of activating cyclic AMP accumulation in slices of these regions of the rat brain (see Table 1), at beta receptors. Furthermore, the actions of norepinephrine and cyclic AMP can be shown to be synergistic in the cerebellum (Fig. 13, 14) and in the hippocampus (Fig. 4). The synergism between responses to norepinephrine and cyclic AMP may indicate that external cyclic AMP can to some extent penetrate into cells sufficiently to lower the thresholds to norepinephrine.

b) Actions of Phosphodiesterase Inhibitors

According to the Sutherland criteria, phosphodiesterase inhibitors should potentiate the effects of transmitters and nerve pathways whose effects are in turn mediated by the postsynaptic formation of cyclic AMP (see III.E). Bearing in mind the precautions of multiple drug actions (see BERKOWITZ et al., 1970) and multiple molecular forms of phosphodiesterase, however, the effects of phosphodiesterase inhibitors have been tested in both the cerebellar Purkinje cells and hippocampal pyramidal cell models. In both cases, the action of norepinephrine and of the locus coeruleus pathway is potentiated by preceding or simultaneous application of the methyl xanthines, aminophylline or theophylline (SIGGINS et al., 1969, 1971; Fig. 15; SEGAL and BLOOM, 1974a) or of papaverine (HOFFER

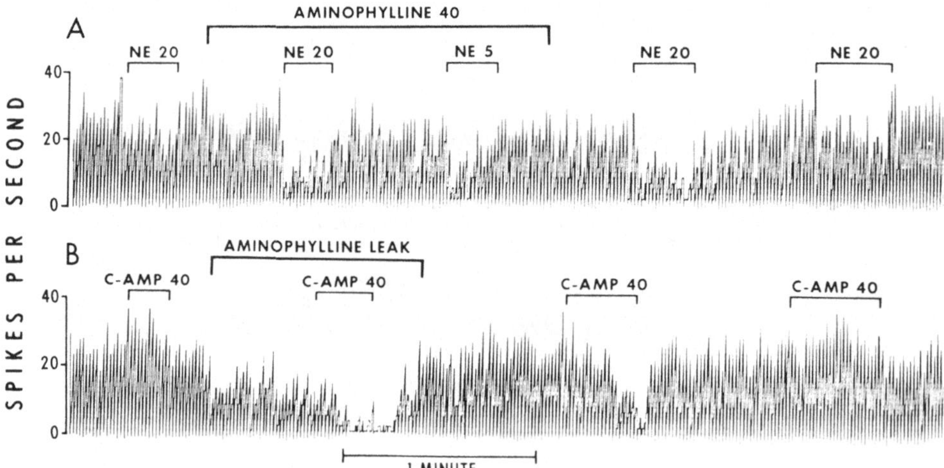

Fig. 15A and B. Potentiation of the responses to NE and cyclic AMP by iontophoretically applied aminophylline. A Slightly effective current of NE (20 nA) becomes almost maximally effective during and shortly after aminophylline (40 nA) is ejected. Previously subthreshold current (5 nA) of NE is now very effective during aminophylline administration. B Same cell as in A, 20 min later. Leakage of aminophylline from micropipet is now sufficient to directly reduce discharge rate. Slight excitation of the unit is noted with 40 nA of cyclic AMP before aminophylline, although during and shortly after aminophylline application, only reduction in firing rate is seen with the same current of cyclic AMP. Recovery of the original excitatory response seen about 2 min after reinstituting aminophylline "holding" current. (From SIGGINS et al., 1971a, with permission of Brain Research)

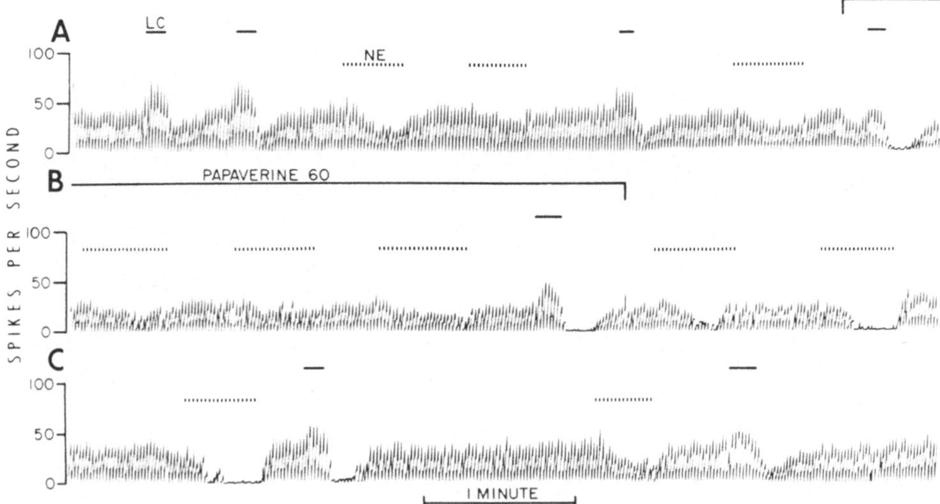

Fig. 16A–C. Potentiation of the effect of norepinephrine and locus coeruleus stimulation by local administration of papaverine: A, B and C represent continuous integrated records from the same cell. Locus coeruleus stimulation (5 V, 0.1 msec square waves, 100 pulses, 10/sec) is denoted by the solid line, and microiontophoresis of NE (80 nA) is represented by the dotted line. Levels of both were adjusted to produce barely threshold response (A). During and shortly after microiontophoresis of papaverine B and C, both NE administration and locus coeruleus stimulation became more effective and produced maximal Purkinje cell inhibition. Several minutes after termination of the papaverine current, NE and locus coeruleus stimulation again produce just threshold changes (C). (From HOFFER et al., 1973: with permission of J. Pharmacol. Exp. Therap.)

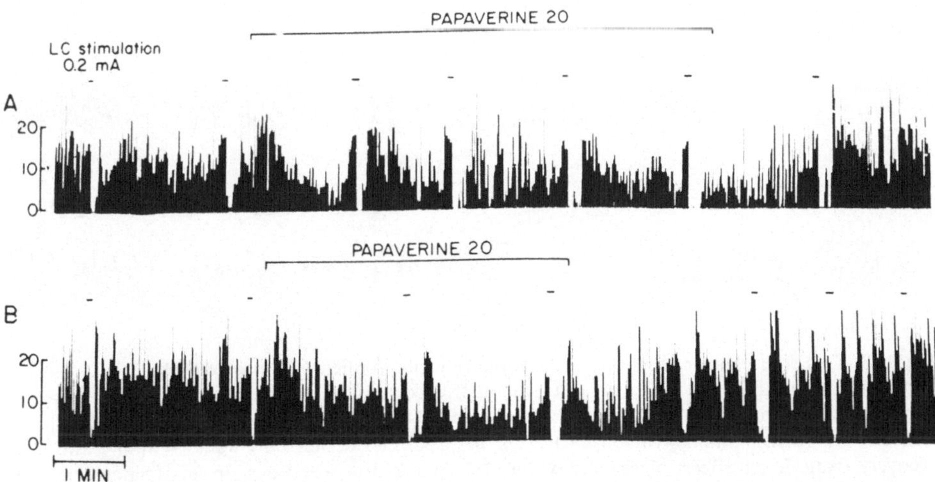

Fig. 17A and B. Papaverine, applied iontophoretically, potentiates the inhibitory actions of the noradrenergic locus coeruleus-hippocampus synaptic pathway. A and B indicate the responses during two different tests. Periods of LC stimulation (100 msec periods of 10 Hz at 0.2 mA) are indicated by the short horizontal bars; periods of papaverine iontophoresis are indicated by the brackets. In both cases, the effects of LC stimulation were potentiated by a 2–3 fold increase in duration with almost immediate recovery. Some direct slowing of the cell by papaverine can also be seen. (From SEGAL and BLOOM, 1974b, with permission of Brain Research)

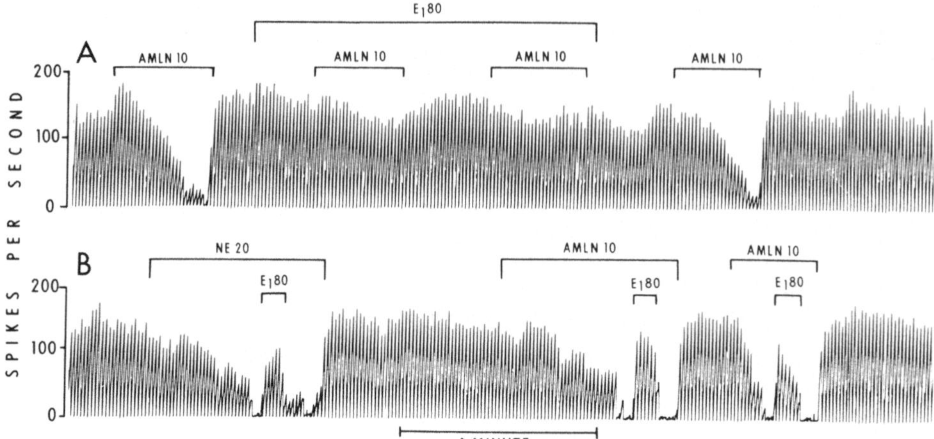

Fig. 18A and B. Blockade by prostaglandin E_1 of depressant responses to NE and aminophylline (AMLN), all applied iontophoretically to a normal Purkinje cell. A Nearly total inhibition of the effects of aminophylline (10 nA) by prior and concurrent application of PGE_1, 80 nA. B Same cell, showing Purkinje cell depressant responses to NE 20 nA and aminophylline 10 nA; during the consequent responses, 80 nA of PGE_1 is ejected, restoring discharge rates to near normal. (From SIGGINS et al., 1971a; with permission of Brain Research)

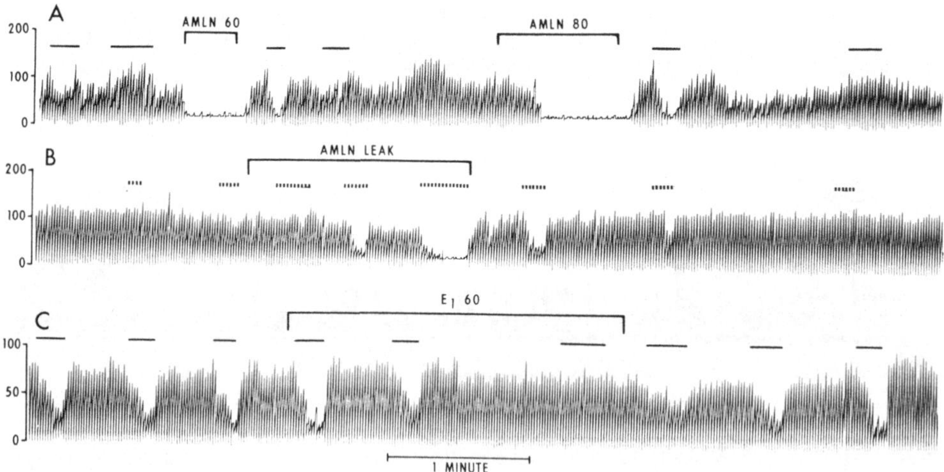

Fig. 19A–C. Effects and interactions of iontophoretically applied NE, aminophylline, cyclic AMP and PGE_1 on Purkinje cells of 6-hydroxydopamine treated animals. A The only Purkinje cell observed which showed a "reversible" or "biphasic" response to NE (solid bars above record indicate application of 80 nA NE). Aminophylline (AMLN) 60 nA abolishes action potentials. Within 1 min after recovery from direct effects of aminophylline, NE 80 nA now abolishes firing. B Same cell as in A. Application of cyclic AMP 200 nA (dashed lines above record) has no effect on firing rate until aminophylline is allowed to leak out of the micropipet. During this time and for 1.5 min after "holding" current is reinstated, cyclic AMP 200 nA now abolishes or slows discharges. C Different cell. Near maximal, reproducible responses to application of NE 80 nA (indicated by solid bars above the record) are blocked by concomitant iontophoretic application of PGE_1 60 nA. Note that the latter is in itself incapable of altering discharge rate. (From SIGGINS et al., 1973; with permission of Brain Research)

et al., 1971 b, 1972, 1973, Fig. 16; SEGAL and BLOOM, 1974a). These drugs also potentiate the action of locus coeruleus stimulation (HOFFER et al., 1973, Fig. 16; SEGAL and BLOOM, 1974, Fig. 17).

Nevertheless, when applied at high iontophoretic currents, aminophylline directly depresses activity (SIGGINS et al., 1971; Fig. 18) even in animals pretreated with 6-hydroxydopamine (Fig. 19). These direct actions of the phosphodiesterase inhibitors have been emphasized in iontophoretic tests on unidentified cerebro-cortical neurons (LAKE et al., 1972, 1973) and in brainstem (ANDERSON et al., 1973 a), where the effects were interpreted as "non-specific" depressant responses. Norepinephrine or locus coeruleus-induced inhibitions can also be potentiated with lower iontophoretic doses of papaverine (HANNA et al., 1972; Figs. 16, 17) or of aminophylline which do not directly depress neurons. Adenosine, known to elevate rat cerebellar cyclic AMP content (see RALL, 1971; DALY, 1975) will

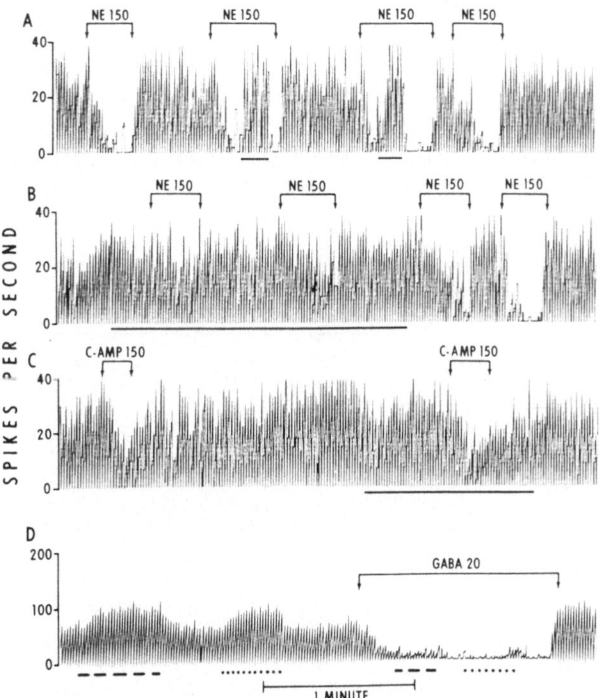

Fig. 20A–D. Effects of microelectrophoretic application of PGE_1 and PGE_2 on spontaneous Purkinje cell discharge. Selective antagonism by prostaglandins of Purkinje cell responses to norepinephrine. A, B, and C Represent consecutive records from the same cell. D illustrated another cell from a different preparation. Duration of NE, cyclic AMP, and γ-aminobutyric acid microiontophoresis indicated by arrows. Numbers after each drug indicate ejection current in nanoamperes. Black lines beneath the records in A, B, and C represent microiontophoresis of PGE_1, 80nA. The dashed and dotted lines beneath the record in D represents microiontophoresis of PGE_1, 125 nA and PGE_2, 125 nA, respectively. Concentrations of prostaglandin which have no direct effect on mean discharge rate (B) antagonize the depressant effects of NE (A, B) but not cyclic AMP (C). Note that doses of prostaglandin which increase mean discharge rates have no effect on the depressant effects of γ-aminobutyric acid (D). (From SIGGINS et al., 1969; with permission of the American Association for the Advancement of Science)

also act non-synergistically with norepinephrine to depress cerebellar Purkinje cells (Fig. 13); this action of adenosine can be potentiated by simultaneous iontophoresis of the phosphodiesterase inhibitor papaverine. On the other hand, effects of adenosine should not be potentiated by methyl xanthine-related phosphodiesterase inhibitors which typically blocks adenosine actions in vitro (SATTIN and RALL, 1970); this latter interaction has not yet been reported for electrophysiologic tests.

c) Pharmacological Blockade of Responses to Norepinephrine

In sympathetic end-organs, some beta receptors can exhibit interesting responses to the prostaglandins: PGE_1 and PGE_2 block the response of adipocytes to norepinephrine, while other prostaglandins show no effects (see ROBISON et al., 1971). In other test cell systems, prostaglandins appear to activate adenylate cyclase. Iontophoretic application of PGE_1 or PGE_2 blocks responses to norepinephrine in cerebellar Purkinje cells (HOFFER et al., 1969; SIGGINS et al., 1971 a, b) and hippocampal pyramidal cells (SEGAL and BLOOM, 1974a; Figs. 18, 19, 20) and also the inhibitions produced in these cells by stimulation of locus coeruleus (Fig. 21; HOFFER et al., 1973; SEGAL and BLOOM, 1974b). As in the responses of the adipocyte, nicotinate will also block the effect of norepinephrine on cerebellar Purkinje cells (Fig. 22; SIGGINS et al., 1971 a).

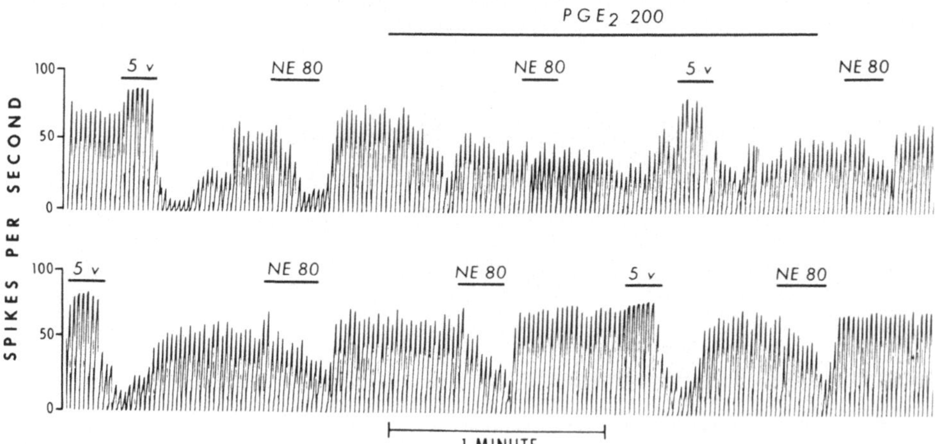

Fig. 21. Effects of PGE_2 on locus coeruleus stimulation and responses to NE in the normal untreated rat. Although the iontophoresis of 200 nA of PGE_2 (long bar above top record) has a direct inhibitory effect on spontaneous activity of this Purkinje cell, it is capable of concurrently blocking the inhibitory effects of both locus coeruleus stimulation (5 V, 50 shocks at 10/sec) and iontophoretically applied norepinephrine (NE, 80 nA of ejection current). The lower record is continuous in time with the top record and shows the rapid recovery of the two responses after termination of the PGE application. The apparent early increase in firing rate occurring during locus coeruleus stimulation (directly below short solid lines labeled 5 v) indicates the addition of the multiphasic stimulus artifact to the "spike counter", and is not a result of synaptic activation by the locus coeruleus stimulation. (From HOFFER et al., 1973; with permission of J. Pharmacol. Exp. Therap.)

In addition, on cerebellar Purkinje cells, the effects of locus coeruleus stimulation and of iontophoretic norepinephrine (but not iontophoretic cyclic AMP) can be inhibited by any of several anti-psychotic phenothiazine derivatives including fluphenazine (Fig. 14), trifluoperazine and alpha flupenthixol (FREEDMAN and HOFFER, 1975; also see V.B.2) but not by the behaviorally inactive phenothiazine derivatives promethazine or beta flupenthixol. Preliminary biochemical studies (SKOLNICK, DALY, FREEDMAN and HOFFER, in preparation) indicate that the psychoactive agents which block the depressant effects of norepinephrine on Purkinje cell discharge also interfere with the norepinephrine-induced accumulation of labeled cyclic AMP in rat cerebellar slices.

d) Actions of Exogenous Cyclic AMP

As discussed above (IV.C.2; Table 4) iontophoretically administered cyclic AMP and dibutyryl cyclic AMP will depress the spontaneous discharge of cerebellar Purkinje cells (Figs. 3, 6, 14, 15, 19, 20; SIGGINS et al., 1969, 1971a; HOFFER et al., 1972, 1973) and hippocampal pyramidal cells (Fig. 4; SEGAL and BLOOM, 1974a). However, in some hands (see GODFRAIND and PUMAIN, 1971; LAKE and JORDAN, 1974) much more variable depressant actions of cyclic AMP have been reported. While technical and biological reasons could account for the reported variation in responsiveness to cyclic AMP (see above and BLOOM et al., 1974), this direct effect or its absence could be considered a crucial piece of evidence. Therefore, two additional types of experiments were undertaken to test the hypothesis that exogenous cyclic AMP causes its effects on norepinephrine target cells by penetrating into the cell to react there with some component such as the cyclic AMP-stimulated protein kinase (III.D.1).

1. Responses to Synthetic Derivatives of Cyclic AMP. If the inhibitory actions of cyclic AMP applied extracellularly are due to effects on an intracellular receptor, derivatives of cyclic AMP which are more liposoluble or more resistant to phosphodiesterase (NEELON and BIRCH, 1973; SIMON et al., 1973; FIKUS et al., 1974) should exhibit greater depressant potency. Although there are exceptions, some halide or acyl substitutions at the 8 position of cyclic AMP (Fig. 1) enhance potency as protein kinase activators in vitro, while N_2 esterifications

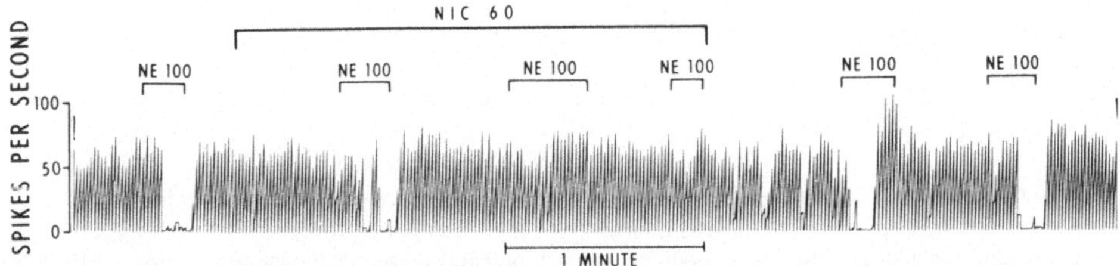

Fig. 22. Inhibitory action of iontophoretically applied nicotinate (NIC, 60 nA) on responses of Purkinje cells to periodic NE (100 nA) administration. Note the total blockade of maximal NE responses by a dose of nicotinate which has no direct action on the spontaneous discharge rate. (From SIGGINS et al., 1971a; with permission of Brain Research)

improve lipid solubility and phosphodiesterase resistivity (SIMON et al., 1973; MEYER and MILLER, 1974). Accordingly, when compared for effects on neuronal activity in vivo, monobutyryl cyclic AMP, 8-p-Cl-phenylthio cyclic AMP, and 8-benzylthio cyclic AMP depress more cerebellar Purkinje cells than does natural cyclic AMP (Fig. 6), and are more potent than cyclic AMP for a given iontophoretic current; conversely, the dibutyryl derivative of cyclic AMP is less effective than cyclic AMP (SIGGINS and HENRIKSEN, 1975). Such results with dibutyryl cyclic AMP do not contradict the known increases in lipid solubility produced by double esterification, but rather, they reflect the decreased potency of dibutyryl cyclic AMP in the protein kinase assay system (FILLER and LITWACK, 1973; NEELON and BIRCH, 1973). Biochemically assessed actions with dibutyryl cyclic AMP, according to the latter results, would be attributed either to contamination with monobutyryl cyclic AMP or to conversion to the monobutyryl form by an endogenous de-acylase (BLECHER and HUNT, 1972). Unfortunately for central neuropharmacological analyses, the latter enzyme is quite low in brain (BLECHER and HUNT, 1972) and this factor could nullify some of the negative results reported with dibutyryl (cyclic AMP).

Thus, presently available data indicate that the relative ability of these cyclic AMP derivatives to activate a bovine brain protein kinase is closely correlated with their ability to depress cerebellar Purkinje cell discharge. Derivatives of cyclic AMP which do not activate the kinase tend to excite (SIGGINS and HENRIKSEN, 1975).

2. Intracellular Recording Experiments. Additional evidence for the possibility that cyclic AMP mediates the responses to norepinephrine synapses in hippocampal pyramidal cells and in cerebellar Purkinje cells is based upon the virtually

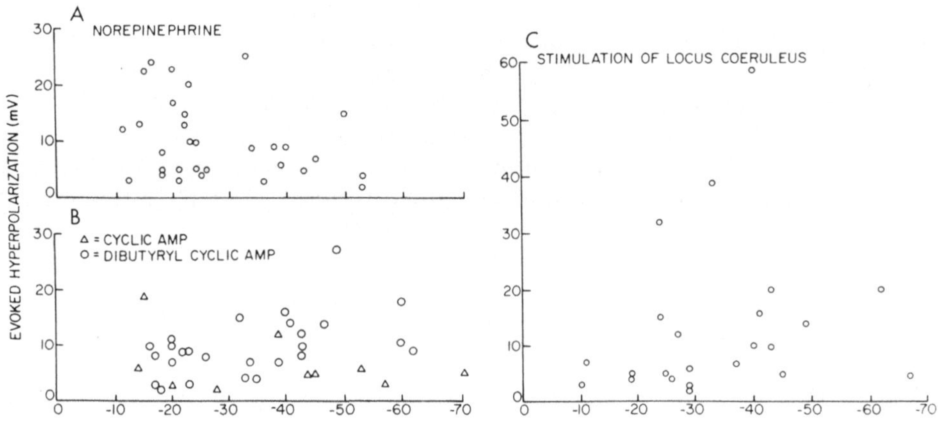

Fig. 23A–C. Relationship of the magnitude of drug or stimulus-evoked hyperpolarizations with initial (resting) membrane potential of individual Purkinje cells. Each point represents the peak hyperpolarizing response for a single test; displacement upward indicates a greater hyperpolarization. A Purkinje cell responses to iontophoresis of norepinephrine. B Responses to iontophoresis of the cyclic nucleotides, cyclic AMP and dibutyryl cyclic AMP. C Responses to stimulation of the locus coeruleus with trains of 100 to 120 pulses at 10/sec. (From HOFFER et al., 1973, with permission of J. Pharmacol. Exp. Therap.)

identical actions of norepinephrine, locus coeruleus-stimulation and iontophoresis of cyclic AMP with respect to the membrane properties of the test cells. All three experimental tests result in hyperpolarization of the target cell with generally increased membrane resistence (SIGGINS et al., 1971, 1972; HOFFER et al., 1973, Figs. 7, 8, 23; OLIVER and SEGAL, 1974). This combination of biophysical changes is not observed with such putative inhibitory transmitters as gamma aminobutyrate or glycine, which hyperpolarize their target cells by decreasing membrane resistance (see WEIGHT, 1971).

If cyclic AMP must penetrate the plasma membrane in order to evoke its response, similar effects should be obtained if it were possible to apply cyclic AMP intracellularly by iontophoresis. Thus on cardiac Purkinje fibers, TSIEN (1973) (also see TSIEN and WEINGART, 1974) has shown that intracellular iontophoresis of cyclic AMP reproduces the actions of cyclic AMP and of norepinephrine applied outside the cell. While such tests require many types of control observations when attempted on central neurons, preliminary results on hippocampal pyramidal cells (OLIVER and SEGAL, 1974) suggest that intracellular iontophoresis of cyclic AMP can also produce both hyperpolarization and increased membrane resistance, as seen after extracellular application.

e) Immunocytochemical Evidence

As described above (Section IV.D1), the immunocytochemical technique for the localization of cyclic AMP indicated that under control conditions only granule cells and a small proportion of Purkinje cells exhibit positive immunoreactivity (BLOOM et al., 1962; SIGGINS et al., 1973; Figs. 9a, b, 10). However, after topical application of norepinephrine in concentrations of 10 μM or more, or during the inhibitory period which follows stimulation of the locus coeruleus in normal rats, a 5–7 fold increase occurs in the number of immunoreactive Purkinje cells (SIGGINS et al., 1973, Fig. 10).

f) Summary

The combined studies on the Purkinje cells of rat cerebellar cortex indicate that norepinephrine elevates cyclic AMP levels in cerebellum and the cyclic AMP immunoreactivity of cerebellar Purkinje cells; that the qualitative changes in discharge rate produced by norepinephrine and of the norepinephrine pathway to both cerebellum and hippocampus can be mimicked by exogenous cyclic AMP and several of its derivatives; that the effects of cyclic AMP on transmembrane potential and ionic conductance changes also mimic the effects of norepinephrine and the pathway on those parameters; that the effects of cyclic AMP, of norepinephrine and of the norepinephrine pathway can be potentiated by phosphodiesterase inhibitors; that the effects of norepinephrine and of the norepinephrine pathway can be antagonized by beta receptor blockers which impair the ability of norepinephrine to activate cyclic AMP accumulation in cerebellar cortex and adipocytes; and that, in preliminary tests, intracellular iontophoresis of cyclic AMP into hippocampal pyramidal cells produces the same hyperpolarization and increase in membrane resistance as does stimulation of the locus coeruleus or iontophoresis of norepinephrine extracellularly. These findings are thus all compatible with the proposal stated earlier that at least for these two test cell systems, the inhibitory synaptic effects of norepinephrine and the cells of the locus coeruleus arise from a sequential response: first they stimulate the formation of intraneuronal cyclic AMP and then the intracellular cyclic AMP activates cyclic AMP-dependent protein kinase to produce a hyperpolarization and increased membrane resistance of long duration.

2. Other Potential Norepinephrine Target Cells and the Effects of Cyclic AMP

Except for the cerebral cortex, responses to norepinephrine or to stimulation of the locus coeruleus in other brain regions have not yet been examined sufficiently to provide a profile with which to compare the effects of exogenous cyclic nucleotides (see BLOOM and HOFFER, 1974; SASA et al., 1974). However, in the cerebral cortex (see Table 4), two types of studies have been done: general surveys of randomly encountered quiescent or spontaneously discharging cells (FREDERICKSON et al., 1972, 1973; LAKE et al., 1972, 1973; JORDAN et al., 1972; PHILLIS et al., 1973; YARBROUGH et al., 1973) and tests of norepinephrine, norepinephrine antagonists, and cyclic AMP and other adenine nucleotides on pyramidal tract neurons identified by antidromic stimulation (STONE et al., 1973; STONE et al., 1975; PHILLIS et al., 1975). In the former case, cyclic AMP did not depress a large proportion of neurons which were depressed by aminophylline or by norepinephrine, but the norepinephrine responses and the aminophylline responses were antagonized by prostaglandins of the E series. In the case of the identified pyramidal tract neurons, almost all cells responded to norepinephrine by slowing of their spontaneous discharge rates; the norepinephrine-induced depressions were antagonized by beta receptor blockers, and more than 75% of the cells responding to norepinephrine were also depressed by cyclic AMP. Despite the good correlation between beta adrenergic receptors and cyclic AMP responses of rat pyramidal tract neurons, a direct anatomical or physiological influence of the locus coeruleus on these neurons remains to be established. It is of interest that slices of guinea pig cerebral cortex which respond to norepinephrine by a 1–4 fold increase in cyclic AMP accumulation (see Table 1) also exhibit under these same conditions an increase in protein phosphorylation which is blocked by added prostaglandin E_1 (WILLIAMS and RODNIGHT, 1974). However, in fetal explant cultures of mouse cerebral cortex and spinal cord (CRAIN and POLLACK, 1973), in neuroblastoma cell lines cultured in high cyclic AMP solutions (CHALAZONITIS and GREENE, 1974) and after topical application of dibutyryl cyclic AMP to the kitten cerebral cortex (PURPURA and SHOFER, 1972) or primary olfactory nerves of the frog (MINOR and SAKINA, 1973), evidence of increased excitability has been reported even though norepinephrine would be expected to be inhibitory in all three regions (see BLOOM and HOFFER, 1974). Such increased excitability could indicate a more pronounced action on membrane resistance by which spike triggering mechanisms are facilitated. On the other hand, in the isolated frog sciatic nerve in vitro, cyclic AMP, but not dibutyryl cyclic AMP, reduces the height of the compound action potential (BERG, 1974).

Attempts to correlate the responsiveness to norepinephrine with the responses to cyclic AMP were also reported by ANDERSON et al. (1973a) in studies on unidentified neurons of the cat brain stem. In this study, cyclic AMP depressed 39 of 41 neurons depressed by norepinephrine, and 34 of 38 neurons depressed by histamine. These cells were also directly depressed by phosphodiesterase inhibitors and the responses to norepinephrine were antagonized by both nicotinate and by prostaglandins of the E series. However, ANDERSON et al. (1973a) interpreted these data as "coincidental" to the hypothesis that the actions of either norepinephrine or histamine on these cells might be mediated by cyclic AMP, because the phosphodiesterase inhibitors theophylline and aminophylline potentiated the effects of the amines in only 3 of 11 trials, and because prostaglandin E_1 blocked only 50% of the depressions produced by norepinephrine. However, in view of the uncertain presence of norepinephrine synapses on any of the test cells, and the possibility of diverse forms of phosphodiesterase isozymes in this region of the cat brain, the data reported by ANDERSON et al. (1973a) would still seem to be in rather good agreement with the other studies detailed above.

C. Dopamine-Mediated Synaptic Events

In the caudate nucleus of the cat (BLOOM et al., 1965; CONNOR, 1970) and of the rat (SIGGINS et al., 1974), the iontophoretic administration of dopamine inhibits firing; these effects of dopamine can be antagonized by some psychoactive phenothiazines (SIGGINS et al., 1974) and mimicked by apomorphine (SIGGINS et al., 1974; Fig. 24). The antidopamine effects of chlorpromazine were demonstrated earlier by YORK (1972) on neurons of the cat putamen. As described above (IV.B1) the activity of a cell free preparation of adenylate cyclase of the rat caudate is stimulated by dopamine and apomorphine, and antagonized by phenothiazines (KEBABIAN et al., 1972; CLEMENT-CORMIER et al., 1974; Table 2). Therefore, caudate nucleus neurons, known to receive dopamine-containing synapses from the substantia nigra pars compacta (CONNOR, 1970; UNGERSTEDT, 1971) clearly fulfill the operational requirements for a test on cyclic nucleotide participation in dopaminergic synaptic function.

The results of such an analysis (SIGGINS et al., 1974) indicate that cyclic AMP and monobutyryl cyclic AMP will depress caudate neurons as do dopamine and apomorphine (Fig. 24). In addition, the depressant effects of dopamine can be potentiated by phosphodiesterase inhibitors; such potentiation is most striking when papaverine and isobutylmethyl xanthine are iontophoretically administered simultaneously (Fig. 25). Recalling the report by ANDERSON et al. (1973a) that methylxanthines did not potentiate the depressant effects of norepinephrine on unidentified neurons of the cat brain stem, it is perhaps pertinent that aminophylline, papaverine, or isobutyl methyl xanthine, when iontophoretically administered alone, would rarely potentiate the effects of dopamine in the caudate nucleus either. However, the combination of inhibitors did produce potentiation (SIGGINS et al., 1974), apparently indicating the specific pharmacological properties of individual caudate phosphodiesterases.

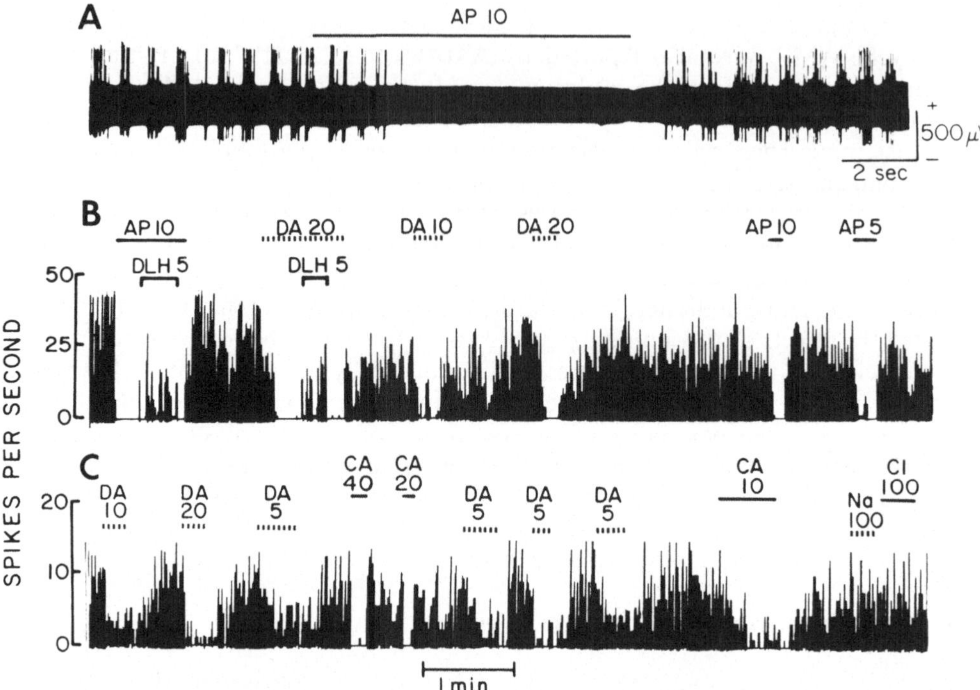

Fig. 24A–C. Effects of microiontophoresis of dopamine (DA), apomorphine (AP) and cyclic AMP (CA) on spontaneous caudate neuronal discharge. A Action potential record filmed from oscilloscope. B and C Polygraph records of firing rate obtained by integration over one-second intervals. Note that application of d,l homocysteate (DLH) restores firing during inhibition by DA or AP, suggesting that excessive depolarization or local anesthetic action is not involved (B). Moreover, large currents of sodium (Na) or chloride (Cl) have little effect on discharge rate (C) indicating that electronic effects of the iontophoretic currents are minimal. These units were recorded in a caudate deprived of its DA nerve terminals by an ascending DA bundle lesion and showed the bursting pattern seen in A. (From SIGGINS et al., 1974; with permission of Pergamon Press)

Unlike the pharmacological effects of norepinephrine in the cerebellum and hippocampus, the catecholamine receptor for dopamine in the caudate was not antagonized either by beta antagonists or by prostaglandins of the E series (SIGGINS et al., 1974).

Dopamine responses, but not cyclic AMP responses, were effectively antagonized by several antipsychotic phenothiazine derivatives, including chlorpromazine, fluphenazine, and alpha flupenthixol (SIGGINS et al., 1974); such studies correlate closely with the actions of these drugs on the dopamine-sensitive adenylate cyclase of caudate (CLEMENT-CORMIER et al., 1974; MILLER et al., 1974; KAROBATH and LEITICH, 1974) and suggest strongly that dopamine transmission in the caudate can be mediated by the cyclic AMP system. In fact, KEBABIAN and GREENGARD have proposed that the caudate adenylate cyclase may well be a model for the dopamine receptor of brain (KEBABIAN et al., 1972). Furthermore, surgical or chemical denervation of the caudate results in supersensitivity

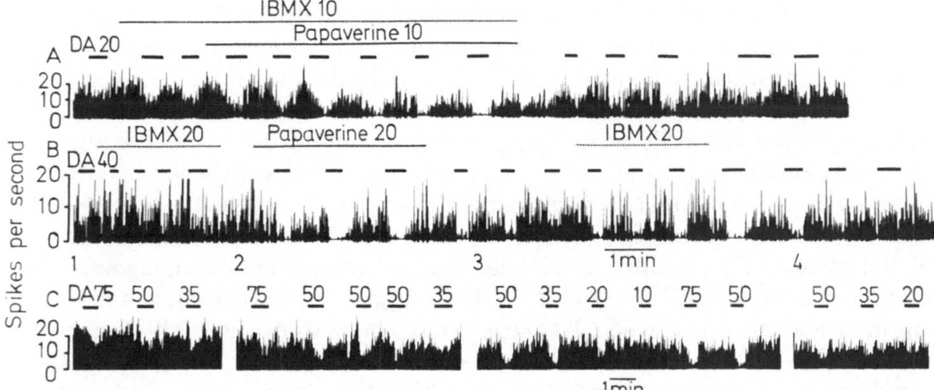

Fig. 25. Effect of phosphodiesterase (PDE) inhibitors on responses of normal rat caudate units to dopamine (DA). solid bars. All cells in this figure were firing spontaneously. Panel A: Concurrent iontophoresis of IBMX (dashed line) and papaverine (solid line). Note that papaverine augments the DA-potentiating effect of IBMX, before papaverine exerts its direct depressant effect. Panel B: prior iontophoresis of papaverine (which has a direct slowing effect) augments the DA-potentiating effect of IBMX. IBMX alone has no direct effect on cell firing rate. Panel C: (1) control responses to DA just before aminophylline injected. (2) 5 min after aminophylline (50 mg/kg) injected intravenously. (3) 5 min after another 50 mg/kg aminophylline injection, and 10 min after end of segment #2 showing potentiation of the DA responses. (4) 50 min after the first aminophylline injection, showing recovery of DA responses to pretreatment levels. (From SIGGINS et al., 1974; with permission of Pergamon Press)

of caudate neuronal responses to dopamine (SIGGINS, HOFFER and UNGERSTEDT, unpublished) as well as increased to dopamine responsiveness of the adenylate cyclase (MISHRA et al., 1974).

The inhibitory effects of iontophoretically administered dopamine have also been compared to the effects of cyclic AMP on a small number of neurons in other brain regions rich in dopamine: the nucleus accumbens and olfactory tubercle (BUNNEY and AGHAJANIAN, 1974) and the entopeduncular nucleus (OBATA and YOSHIDA, 1973). In both reports, cyclic AMP depressed neuronal discharge as did dopamine. The physiologic significance of dopaminergic fibers within the cerebral cortex remains to be pursued (BERGER et al., 1974; FUXE et al., 1974; LINDVALL et al., 1974).

D. Sympathetic Ganglia and the Synaptic Roles of Catecholamines and Cyclic Nucleotides

Due to their relatively simple cellular composition, sympathetic ganglia have been a favorite object for the analysis of synaptic transmission (see ECCLES, 1964; and reviews by WEIGHT, 1971, 1974; GREENGARD and KEBABIAN, 1974). Originally, these ganglia were considered as simple nicotinic, cholinergic relays between presynaptic neurons of the spinal cord and the adrenergic post-ganglionic neurons. However, most recent studies have concentrated on the proposal of

ECCLES and LIBET (1961) that chromaffin cells in the sympathetic ganglia might function as inhibitory adrenergic interneurons, being excited by a collateral preganglionic muscarinic cholinergic receptor and producing on the post-ganglionic neurons a slow postsynaptic inhibitory potential. Subsequent reports indicated that the chromaffin cells of rat ganglia might contain dopamine (NORBERG et al., 1966; BJÖRKLUND et al., 1970). Furthermore, these cells were described by ERANKÖ and HARKONEN (1965) as "small, intensely, fluorescent" or S.I.F. cells, and seemed to provide the structural basis for the pharmacological analysis of the effects of exogenous catecholamines or catecholamine antagonists (see LIBET and KOBAYASHI, 1969; LIBET, 1970) in support of the catecholamine-secreting interneuron. Thus, when KEBABIAN and GREENGARD later (1971) reported that dopamine could increase the cyclic AMP content of bovine sympathetic ganglion slices, this result was linked to the previously reported observation that synaptic activation of the ganglia would also increase ganglionic cyclic AMP content in the rabbit (MCAFEE et al., 1971) to suggest that cyclic AMP could be involved in the catecholaminergic synaptic actions of these interneurons (see GREENGARD and KEBABIAN, 1974).

The following observations provide a strong case for the possible role of cyclic nucleotides in sympathetic ganglionic transmission: in the intact rabbit ganglion, synaptic activation in vitro results in increased cyclic AMP content (MCAFEE et al., 1971) as does dopamine in the slice preparation of bovine ganglia (KEBABIAN and GREENGARD, 1971).

Norepinephrine and histamine, but not dopamine, increase cyclic AMP in the cultured rat cervical ganglion (CRAMER et al., 1973; LINDL and CRAMER, 1974), as do muscarinic cholinergic agonists, in rabbit ganglia. The latter effect does not occur in the absence of Ca^{++} (KALIX et al., 1974) suggesting that it is indirectly evoked through the release of a second transmitter substance. Furthermore, the increased levels of cyclic AMP produced in the cow and rabbit ganglia can be antagonized only by cholinergic muscarinic blockers or by alpha adrenergic blockers (KALIX et al., 1974). In addition, dopamine and cyclic AMP will hyperpolarize post-ganglionic neurons of the rabbit and the effects of dopamine are potentiated by theophylline and antagonized by prostaglandin E_1 (MCAFEE and GREENGARD, 1972; KALIX et al., 1974; Fig. 26). The elevation of cyclic AMP produced in bovine ganglion slices by dopamine can also be localized to the post-ganglionic neurons by immunocytochemistry (Fig. 9C; KEBABIAN, BLOOM, STEINER and GREENGARD, unpublished results).

Exposure of bovine ganglia slices to acetylcholine will selectively increase cyclic GMP content, an effect which is also sensitive to muscarinic antagonists (GREENGARD and KEBABIAN, 1974) and which is, furthermore, localized to post-ganglionic neurons (KEBABIAN, BLOOM, STEINER and GREENGARD, unpublished). Cyclic GMP depolarizes rabbit ganglion neurons (MCAFEE and GREENGARD, 1972), an effect which is similar to the slow cholinergic muscarinic excitatory potential seen in these ganglia. Recently, synaptic stimulation of the frog sympatheic ganglia was reported to produce a prompt elevation of the ganglionic cyclic GMP content within the period of stimulation required to generate the slow excitatory cholinergic potential in the frog (WEIGHT et al., 1974); synaptic stimulation of the frog ganglia also increased cyclic AMP content (WEIGHT et al., 1974).

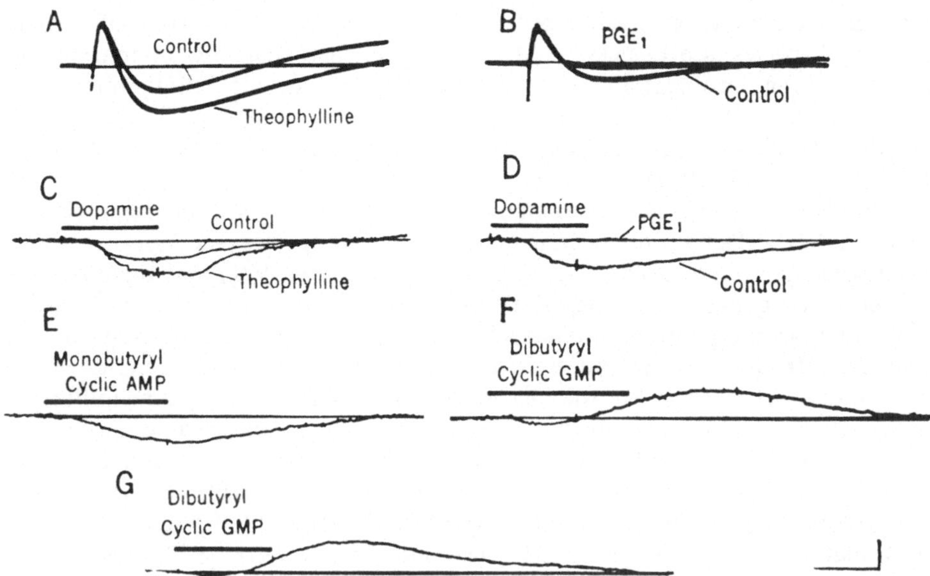

Fig. 26A–G. Effect of dopamine and of cyclic nucleotides and of agents which affect cyclic AMP metabolism on synaptic and resting membrane potentials recorded from postganglionic neurons of the rabbit superior cervical ganglion by means of the sucrose gap technique. A. and B. Oscillographic traces of electronically conducted synaptic potentials elicited in response to a single supramaximum stimulus to the preganglionic nerve. Hexamethonium chloride (600 μM) was present to abolish propagated responses. A. Responses obtained in Locke solution and after 30 min of superfusion with Locke solution containing 1.5 mM theophylline are superimposed. B. Responses obtained in normal Locke solution and after 15 min of superfusion with Locke solution containing 3×10^{-7} M PGE_1 are superimposed. Results similar to those illustrated here were obtained when d-tubocurarine (125 μM) was used instead of hexamethonium chloride to abolish propagated responses. This dose of d-tubocurarine also abolished the initial EPSP. C. and D. Resting membrane potential changes in response to a brief period of superfusion with dopamine. C. Responses to 50 μM dopamine before (control) and 30 min after the start of superfusion with 2 mM theophylline are superimposed. D. Responses to 200 μM dopamine before (control) and 20 min after the start of superfusion with 6×10^{-7} M PGE_1 are superimposed. E. and F. Changes in membrane potential in response to a brief period of superfusion with 2.5 mM monobutyryl cyclic AMP (E) or 25 μM dibutyryl cyclic GMP (F). G Change in membrane potential of the cervical vagus nerve in response to a brief period of superfusion with 200 μM dibutyryl cyclic GMP. The duration of superfusion with Locke solutions containing dopamine or cyclic nucleotides is indicated by the solid bars. All records d-c recording, hyperpolarization downward. Calibration marks: (A and B) 1 sec, 800 μv; (C through G) 2 min, 400 μv. (From McAfee and Greengard, 1972; with permission of the American Association for the Advancement of Science)

As in the case of the many biochemical studies of transmitters and resultant cyclic nucleotide dynamics described earlier, it is often possible to generalize results across species lines, although the magnitude and occasionally the qualitative effects of a given transmitter may show marked species variation (see Daly, 1975). Similarly, species considerations also play an important role in the interpretation of the data relating catecholamines and cyclic nucleotides to the generation of synaptic potentials in sympathetic ganglia. On the one hand, it seems clear that the electrical waves or after potentials recorded from the surface of ganglia of several species (see ECCLES, 1964) can be related to the slow inhibitory and excitatory synaptic potentials recorded with sucrose gap or intracellular microelectrodes (LIBET, 1970; WEIGHT and PADJEN, 1973a, b; WEIGHT and VOTAVA, 1970), and that SIF-like cells can also be seen in ganglia of many species (see JACOBOWITZ, 1970; BLACK et al., 1974).

On the other hand, a detailed analysis of the ultrastructural interconnections between these cells has only been achieved for a few species (see YOKOTA, 1973; WILLIAMS and PALAY, 1969; TAMARIND and QUILLIAM, 1971) and therefore the existence of a common cytological arrangement accounting for the presumptive dopaminergic interneuronal linkage still requires additional documentation. For example, SIF-like cells are known to occur in the sympathetic ganglia of the frog (JACOBOWITZ, 1970) but the pharmacological analysis of the slow inhibitory potential in the frog suggests that it is the result of a muscarinic cholinergic transmission (WEIGHT and PADJEN, 1973a, b). Furthermore, in the superior cervical ganglion of the rabbit, the effects of exogenous dopamine have been interpreted as resulting in presynaptic inhibition (DUN and NISHI, 1974), rather than the postsynaptic hyperpolarization (MCAFEE and GREENGARD, 1972) observed in rabbit ganglia exposed to dopamine or to cyclic AMP (Fig. 26). Although DUN and NISHI (1974) also observed a hyperpolarizing action of dopamine on post-ganglionic neurons, the reduced effectiveness of pre-ganglionic excitation in the presence of normal responses to applied acetylcholine during this dopamine action indicated to these investigators that the primary site of dopamine action was on the release of presynaptic transmitter; however, this action alone would not seem sufficient to explain the hyperpolarization induced by dopamine. Although dopamine and cyclic AMP both result in hyperpolarizing responses, even higher concentrations of dopamine do not produce substantial elevations of whole rabbit ganglionic cyclic AMP under the same conditions (KALIX et al., 1974).

Overall, however, the studies summarized above suggest that cyclic nucleotide mediation of sympathetic ganglionic transmission might account for both the excitatory (cyclic GMP) and inhibitory (cyclic AMP) effects which accompany tetanic pre-ganglionic stimulation. Even so, some points of transmitter identification, their cellular basis, and ionic mechanisms of action remain to be worked out in greater detail (see GREENGARD and KEBABIAN, 1974; and KUBA and KOKETSU, 1974).

E. Biophysical Parameters of the Effects of Catecholamines in Other Cells Relevant to Cyclic Nucleotides

When the changes in neuronal membrane properties produced by a nerve pathway are shown to be identical with the effects of a cyclic nucleotide on these target neurons, this similarity of action suggests that cyclic nucleotides might mediate intracellularly the transmission across this synapse. In the examples reviewed above, hyperpolarizing responses associated with decreased ionic conductance have been observed in response to norepinephrine and cyclic AMP in the cerebellum and hippocampus and, in addition, hyperpolarizing responses were also observed with dopamine and cyclic AMP in the sympathetic ganglion.

Since the association of a hyperpolarizing response with decreased ionic conductance represents a mode of response which differs from the more classical synaptic potentials associated with increased ionic conductances, it seems pertinent to examine briefly the effects caused by catecholamines in other tissues in which these biophysical parameters have been measured.

In several types of skeletal muscle, catecholamine receptors characterized pharmacologically as beta have been reported to result in hyperpolarization associated with decreased membrane conductance (HIDAKA and KURIYAMA, 1969; KUBA, 1970; KUBA and TOMITA, 1971; ITO et al., 1971). Interestingly, catecholamines will also affect the motor nerves of these same preparations, but generally these responses have been characterized pharmacologically as alpha, restoring junctional transmission (BRECKENRIDGE et al., 1967) and facilitating the frequency of miniature endplate potentials (JENKINSON et al., 1968; GOLDBERG and SINGER, 1970; KUBA, 1970; KUBA and TOMITA, 1971). In several of these studies the effects of cyclic AMP were similar to those of the catecholamine on presynaptic function (GOLDBERG and SINGER, 1970; TAKAMORI et al., 1973).

In cardiac muscle (TSIEN et al., 1972; TSIEN, 1973) catecholamines act at a beta receptor, leading to increased membrane conductance and depolarization; similar changes have been described for catecholamine receptors of the guinea pig taenia coli (BÜLBRING and TOMITA, 1969; MAGARIBUCHI and KURIYAMA, 1972; TOMITA et al., 1974) and for oviduct (see BRUNTON, 1972). On the other hand, hyperpolarizing catecholamine receptors in the smooth muscle of blood vessels (SOMLYO et al., 1970) and uterus (KROEGER and MARSHALL, 1973; VESIN and HARBON, 1974) are associated with increased potassium conductance, as are the hyperpolarizing responses of hepatocytes to epinephrine (HAYLETT and JENKINSON, 1969; FRIEDMANN et al., 1971; PETERSEN, 1974). In the pineal (SAKAI and MARKS, 1972; KAKIUCHI and MARKS, 1972) and in brown fat (HORWITZ et al., 1969; HOROWITZ et al., 1971; KRISHNA et al., 1970) beta adrenergic receptors have also been examined for biophysical actions: in the pineal, hyperpolarizing responses were attributed to increased potassium conductance, while in brown fat the depolarizing response is associated with increased conductance for both sodium and potassium. On Aplysia neurons, hyperpolarizations due to catecholamines have also been observed with either decreased or increased resistances (ASCHER, 1974; CARPENTER and GAUBATZ, 1974). On the other hand, in the mammalian spinal cord, hyperpolarizing responses to norepinephrine are associated with decreased conductance (ENGBERG and MARSHALL, 1970; ENGBERG et al., 1974).

The tentative conclusion from this broad ranging but superficial resume of multiple cell types must be that the ionic mechanisms of catecholamine receptors cannot easily be generalized. While it seems highly likely that cyclic AMP fully replicates the action of adrenaline on heart muscle (TSIEN et al., 1972; BROOKER, 1974) even when it is introduced directly into the cardiac Purkinje fiber (TSIEN, 1973; TSIEN and WEINGART, 1974), it also seems clear that other actions of catecholamines cannot now be fully satisfied by this explanation (see DIAMOND, 1973). As in the case of the serotonin receptors of insect salivary glands (BERRIDGE and PRINCE, 1971, 1972) and of adrenergic receptors of rat salivary gland (BATZRI et al., 1971), more than one type of membrane change may accompany the response to a given transmitter especially when a secretory action is induced. Thus, an even wider variety of biophysical changes might be expected from analysis of the effects of cyclic AMP on the guinea pig submandibular gland which exhibits different secretory patterns to norepinephrine, serotonin, and to cyclic AMP (CARLSOO et al., 1974).

F. Are All Catecholamine-Triggered Events Mediated by Cyclic AMP?

Data from some synapses, reviewed above, support the concept that catecholamines evoke functional changes from certain postsynaptic cells by regulating cyclic AMP levels which in turn may result in the phosphorylation or dephosphorylation of membrane proteins or other cell components. If this concept holds for some synapses mediated by these transmitters, an important related consideration is whether all synapses mediated by these transmitters would also be expected to function by the same mechanism. However, all junctions in which transmission may be mediated by the secretion of catecholamines clearly do not seem to be identical: thus, we have specific patterns of pharmacological interaction with alpha or beta antagonists or with prostaglandins which in part serve to separate two classes of norepinephrine receptors from one class of dopamine receptors; at the same time some drugs like the antipsychotic phenothiazines work on both classes of catecholamine receptors. Furthermore, even within the cells characterized as having beta receptors in the peripheral nervous system, the ionic basis of the activation of these receptors varies substantially; only for cardiac muscle, liver and adipocytes are there strong indications that cyclic AMP mediates these catecholamine actions.

Because of the variations in molecular mechanisms for the catecholamine receptors which have been resolved, there does not now seem to be any obvious simple method by which to predict whether or not cyclic nucleotides may be involved. Conversely, the potential technical shortcomings in the biochemical and electrophysiological assessment of cyclic nucleotide functions also prevent the conclusion that no changes in cyclic nucleotide level accompany the response of some cells in a given brain region to one or more transmitter substances. This is because so few transmitter substances have now been tentatively identified that the combination of the right substance with the proper target cells may not yet have been achieved; furthermore, the optimal conditions for assessing the biochemical and physiological actions of natural and synthetic nucleotides undergo almost daily evolution. In the light of these considerations, therefore, the question of whether all catecholamine events are mediated by cyclic AMP cannot be generalized beyond those synaptic junctions where all the criteria of transmitter-identification and second messenger mediation have been fully probed.

G. Catecholamines and the Role of Ca^{++}

The mediation of the actions of catecholamines (and other transmitters or hormones) may well involve fundamental intracellular chemical changes in addition to those of the cyclic nucleotides. In particular, changes in intracellular and transmembrane Ca^{++} appear to be crucial. When considered as a cellular regulator in its own right, Ca^{++} clearly functions in multiple roles within innervated tissues, including maintenance of membrane integrity, maintenance of intercellular junctions, spike generation, excitation-secretion coupling, excitation-contraction coupling, and regulation of the activity of many enzymes (see BIANCHI, 1968; TRIGGLE, 1971).

Originally, none of the mechanisms which were thought to control Ca^{++} fluxes and Ca^{++} binding within cells were thought to involve cyclic nucleotides. More recent reports indicate, however, that cyclic AMP can in fact initiate the sequestration of Ca^{++} by endoplasmic reticular fragments in skeletal, cardiac and smooth muscle (KATZ et al., 1974; TADA et al., 1974; KIRCHBERGER et al., 1974; BROOKER, 1974; WATANABE and BESCH, 1974; BLOOM and SWEAT, 1974; DHALLA et al., 1973; MITZNEGG et al., 1974; CHIARANDINI et al., 1973; KADLEC et al., 1973; KATZ and TENENHOUSE, 1973; VERMA and MCNEILL, 1974; VESIN and HARBON, 1974; WILL et al., 1973). Furthermore, other data (IIIA–D, above) indicate that while Ca^{++} is essential for the basal activity of adenylate and guanylate cyclase, higher levels of Ca^{++} can alter the responsiveness of the enzyme activity to hormones or transmitters. Furthermore, Ca^{++} can activate at least one isozyme of phosphodiesterase, and changes in the levels of Ca^{++} within the biological range (i.e., micromolar) can reverse the effects of cyclic AMP on phosphotransferase reactions from activation to inhibition (UEDA et al., 1973) for at least some synaptic membrane protein substrates. In addition to these interactions, RASMUSSEN (1970) has suggested that any process which drew upon ATP stores, such as adenylate cyclase activation, might well liberate intracellular Ca^{++} simultaneously.

Therefore, the significant question arises whether catecholamines produce their effects by activating adenylate cyclase (and thereby initiating intracellular Ca^{++} translocation and a sequence of Ca^{++}-sensitive events to execute the catecholamine-elicited cell response), or whether the active catecholamine receptor produces first an increased Ca^{++} conductance, which could then change several intracellular enzyme activities (see DIAMOND, 1973), including nucleotide cyclases, phosphodiesterase and phosphotransferases (see Section III.D.3). These two parallel conceptualizations are diagrammed in Fig. 27, although an eclectic interpretation such as in the response of the insect salivary gland to serotonin (BERRIDGE and PRINCE, 1971, 1972), indicates that the two processes may be simultaneous rather than mutually sequential (also see RASMUSSEN, 1970).

Some iontophoretic experiments have been interpreted to indicate that the effects of all inhibitory monoamines on unidentified cortical neurons were mediated by changes in Ca^{++} flux and not by cyclic nucleotides (see PHILLIS et al., 1973; PHILLIS, 1974a; LAKE et al., 1972, 1973; YARBROUGH et al., 1974). In these experiments, cells which responded to norepinephrine or serotonin by depression of discharge appeared to lose this response when the cells were exposed

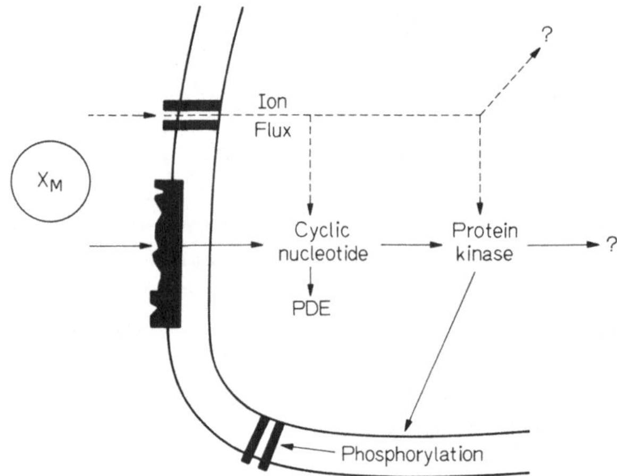

Fig. 27. Schematic representation of two possible sequences by which the ionic fluxes across the postsynaptic membrane may participate in the response to neurotransmitter substances (Xm) and cyclic nucleotides. In one view, the transmitter acts to alter membrane channels limiting ionic movements; changes in this flux could modulate both cyclic nucleotide synthesis and protein kinase as well as other intracellular events including membrane properties. Alternatively, the transmitter could first activate the cyclic nucleotide synthesis, which then, through protein kinase intermediates and possible phosphotransferase reactions involving membrane proteins, alters the permeability of the membrane. Eclectic interpretations are also possible (see text; drawn by Dr. Ante Padjen)

to the iontophoretic administration of the presumed Ca^{++} antagonist verapamil (see CHIARANDINI and BENTLEY, 1970), or the Ca^{++} displacing ions, Mn^{++} and La^{++} (see GOODMAN and WEISS, 1971). These data could indicate that the loss of responsiveness to the monoamines resulted from the displacement of Ca^{++} from essential membrane ligand or receptor site. The same data could also simply indicate that the loss of responsiveness was more apparent than real, and related to the prolonged depressant actions of the divalent metals presumed to be Ca^{++} antagonists (ROZEAR et al., 1971). If, because of this metal depressant effect, the iontophoresis of monoamines were interrupted until spontaneous or amino acid-induced activity recovered, then reduced or "absent" responses to the first few applications of the monoamine could result directly from failure to overcome the iontophoretic retention currents (see BRADSHAW et al., 1973; FREEDMAN and HOFFER, in preparation). On cerebellar Purkinje cells (Fig. 28), Ca^{++}, like cyclic AMP, is generally inhibitory, while Ca^{++}-chelating agents, including ATP, produce transient excitatory effects (SIGGINS et al., 1971 a).

In any event, judicious pursuit of the complicated interactions between cyclic nucleotide-regulated events and Ca^{++}-regulated events within neurons and other cells seems clearly warranted. For example, if the initial synaptic action of catecholamines were to be the enhancement of transmembrane Ca^{++} movements, a net influx—such as accompanies the process of excitation-secretion coupling (see above)—would require a depolarizing response with increased membrane conductance. In fact, however, just the opposite effects of catecholamines on

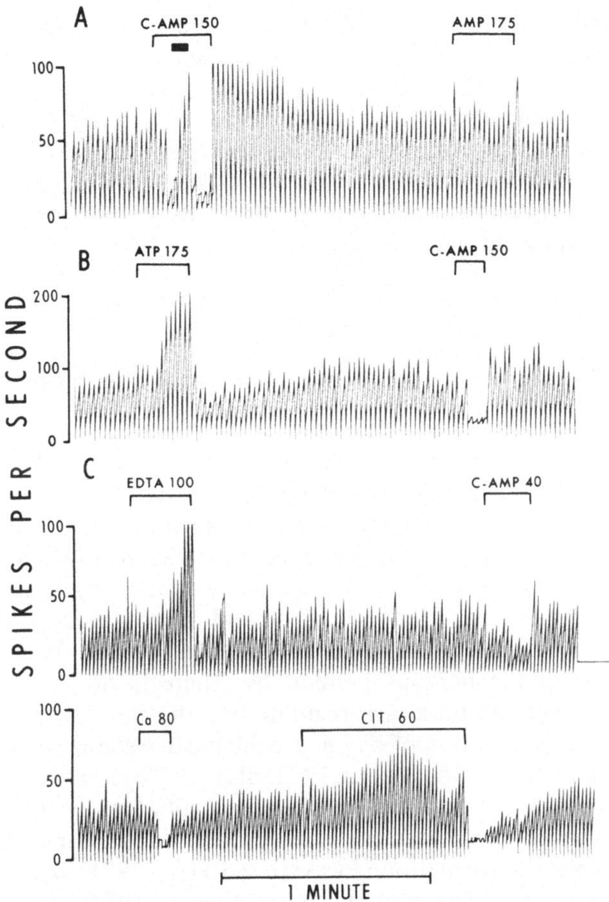

Fig. 28 A–D. Influence of iontophoretically applied adenine nucleotides, calcium chelators and Ca^{2+} on spontaneous discharge of Purkinje cells. A and B Taken from same cell. Heavy bar during cyclic AMP administration in A shows cathodal stimulation of the cell through another (NaCl) barrel of the 5-barrel pipet. Increased discharge during this stimulation suggests that the cyclic AMP depression is not the result of local anesthesia or hyperdepolarization. Note that while cyclic AMP only depresses activity in these cells 5′-AMP does nothing (A), and ATP (B), EDTA (C), and citrate (D), all accelerate firing. Calcium always slowed activity (D). Often, supramaximal currents of ATP and the chelators accelerated discharge presumably to the point of "depolarization block". (From SIGGINS et al., 1971a; with permission of Brain Research)

membrane potential and membrane conductance are observed with intracellular recordings (SIGGINS et al., 1971c, d; ENGBERG and MARSHALL, 1971, ENGBERG et al., 1974). The hyperpolarization observed in these latter studies could theoretically have arisen from an increased Ca^{++} efflux, but if this were the primary event in the hormonal response (see DAMBACH and FRIEDMANN, 1974; FRIEDMANN et al., 1971; SOMLYO et al., 1970; MATTHEWS and SAFFRAN, 1973), conductance should—at least initially—have been increased, and extracellular Ca^{++} ought to block the ionic efflux and reverse the effect, rather than inhibiting discharge activity directly. Alternatively, the effects of La^{++} in displacing Ca^{++} from an essential intracellular locus such as the adenylate or guanylate cyclase or a subsequent phosphotransferase, could block a monoamine response by interrupting the intracellular cyclic nucleotide reaction sequence leading to the sets of changes which result in hyperpolarization (also see WILL et al., 1973). In cardiac muscle (TSIEN, 1973; BORASSIO and VASSALE, 1974; VASSALE and BARNABEL, 1971; MEINERTZ et al., 1974), potassium ions seem to be the major species affected by norepinephrine.

VI. Cyclic Nucleotides and Other Neurobiological Functions

In studies on many types of non-neuronal cells, a substantial body of evidence suggests that cyclic nucleotide-sensitive processes may be able to regulate a very wide range of biological phenomena including such events as cell division (FROELICH and RACHMELER, 1972; WHITFIELD et al., 1973; MCMANUS and WHITFIELD, 1974; PAPAHADJOPOULOS et al., 1974; also see VOADEN, 1971), secretion (HEISLER et al., 1972; MATTHEWS and SAFRAN, 1973) and microtubular assembly (MAGUN, 1973; EIPPER, 1974). All of these same biological events are also fundamental to the function of the nervous system, but there is no indication at present whether such events as these are regulated in the developing nervous system by the sequential maturation of key neurochemical systems such as the catecholamines (BLOOM et al., 1974b; LENTZ, 1972; also see FUXE et al., 1974b). Although an extended correlation may not yet be fully justifiable, cyclic nucleotides have already been shown to influence cultured neurons and glia in such morphological parameters as neurite formation (PRASAD and HSIE, 1971) when grown on glass coverslips, but not on collagen (MILLER and LEVINE, 1972), on axonal elongation (ROISEN et al., 1972a, b, c; ROISEN and MURPHY, 1973; SHEPPARD and PRASAD, 1973), the morphology and nucleic acid synthesis of glia (GIBSON et al., 1974; EDSTROM et al., 1974; SHEPPARD et al., 1975), and generalized trophic effects on developing neuromuscular junctions (LENTZ, 1972).

The functional significance of these trophic actions of exogenous cyclic nucleotides in vitro for the developing nervous system in vivo have thus far not been exclusively associated with any particular transmitter substance (see SCHMIDT and ROBISON, 1971; SCHMIDT et al., 1971; VON HUNGEN and ROBERTS, 1974). It is of interest, however, that dopamine shows decreased ability to activate adenylate cyclase in the senscent rat caudate (WALKER and WALKER, 1973b; also see WILLIAMS and THOMPSON, 1973). Furthermore, although dopamine activates adenylate cyclase in bovine and rat retina (BROWN and MAKMAN, 1972, 1973), a tissue rich in phosphodiesterase (PANNBACHER et al., 1972, 1973), the uncertain cellular location of dopamine-secreting retinal neurons and the unknown functional consequences of this effect (BENSINGER et al., 1974) prevents any correlation yet with visual reception (MILLER, 1973). However, an ATP-dependent phosphodiesterase with substrate specificity for cyclic GMP has been reported to be activated by brief photic stimulation (MIKI et al., 1973; BITENSKY et al., 1971, 1973). Although these observations are functionally provocative, additional observations will be required to establish their precise biological significance.

A. Cyclic Nucleotides and Behavior

At the behavioral pole of the neurobiological spectrum, intriguing associations have also been reported which suggest that the ability of cyclic nucleotides to influence nucleic acid, ribosomal and protein metabolism in neurons (see SHASHOUA, 1971; GIBSON et al., 1974; WILLIAMS and RODNIGHT, 1974; SCHMIDT and SOKOLOFF, 1973; LEVITAN et al., 1974) may be indicative of a role of cyclic nucleotides in the establishment of longer term behavioral events. Several studies on whole animal behavior have suggested that cyclic nucleotides might be able to influence a wide range of behavioral responses (KRISHNA et al., 1968; CHOU et al., 1971; CONTRERAS et al., 1972; HERMAN, 1973; COHN et al., 1974; WILLIAMS and PIRCH, 1974), including opiate dependence and analgesia (HO et al., 1973a, b; NAITO and KURIYAMA, 1973) and some actions of ethanol (VOLICER and GOLD, 1973; KURIYAMA and ISRAEL, 1973), as well as possible human affective disorders (ABDULLAH and HAMADAH, 1970; PAUL et al., 1971a, b; CRAMER et al., 1972; MURPHY et al., 1973). In addition, BRECKENRIDGE and LISK (1969) reported that intra-hypothalamic cyclic AMP pellets can block the lordotic posture normally induced in ovariectomized rats after progesterone and suggested that cyclic nucleotides could regulate some neuroendocrine functions. In fact, estradiol will regulate hypothalamic adenylate cyclase in vitro (GUNAGA and MENON, 1973) but this effect may be indirect since it is antagonized by adrenolytic agents. Since several actions of estradiol on uterine function (MITZNEGG et al., 1974; VESIN and HARBON, 1974; VOKAER et al., 1974) may also be associated with cyclic nucleotide-mediated events, these preliminary findings in hypothalamus take on possible functional significance. In addition, cyclic nucleotides in the hypothalamus can alter learned feeding behaviors (BOOTH, 1973), and cyclic nucleotide-induced changes would be presumed to follow the intrahypothalamic injection of minute amounts of cholera toxin (CLARK et al., 1974). Thus it will be of interest to determine whether any other longer-term events, such as the activation of some catecholamine-containing neurons during sleep (CHU and BLOOM, 1973, 1974a, b; also see JOUVET, 1972) are correlated with cyclic nucleotide alterations.

Furthermore, evidence cited earlier (see V.C; Table 2) has suggested behavioral significance for a dopamine-sensitive adenylate cyclase receptor model developed from cell-free homogenates of the caudate nucleus (KEBABIAN et al., 1972; CLEMENT-CORMIER et al., 1974). In these and other biochemical studies of dopamine on caudate homogenates (KAROBATH and LEITITCH, 1974; MISHRA et al., 1974; MILLER et al., 1974; SHEPPARD and BURGHARDT, 1974) and homogenates of other dopamine-rich regions of the limbic forebrain (MILLER et al., 1974) the cyclase stimulation by dopamine is antagonized by phenothiazines and butyrophenones in a close correlation to their rank ordering as clinically effective antipsychotic agents. While such evidence may not establish that a dopamine-adenylate cyclase hyperactivation underlies the chemical pathology of psychosis, the predictive value of this receptor model may well provide a more sound basis on which to develop potent specific anti-psychotic agents (see KEBABIAN et al., 1972; MILLER et al., 1974). For completeness sake, however, it must be noted that anti-psychotic phenothiazines have many actions. For example, these drugs also impair the release of dopamine from brain slices in vitro (SEEMAN and LEE, 1974) again in a rank order potency even more closely approximating clinical effectiveness. Furthermore, phenothiazines can also impair the activation of cyclic AMP accumulation produced in brain slices by norepinephrine (BLUMBERG and SULSER, 1974; PALMER et al., 1972a, b; SKOLNICK and DALY, personal communication) as well as blocking the response to norepinephrine of cerebellar Purkinje cells (FREEDMAN and HOFFER, 1975). Similarly, these drugs—in substantially higher concentrations—are known to inhibit one or more phosphodiesterase isozymes (see UZUNOV and WEISS, 1972) and to anesthetize electrically excitable membranes. Clearly, a definitive answer has not yet been obtained as to which components of the central catecholamine synaptic pathways or their interrelated cyclic AMP system represent the site of the therapeutic action of these anti-psychotic compounds, if any.

B. Cyclic Nucleotides and Enzyme Activation

In a number of animal and bacterial cells, longterm alterations in intracellular cyclic nucleotide content lead to the induction of various enzymes essential for the function of the differentiated cell (see LANGAN, 1973; WICKS, 1974). It is not yet clear whether the changes in cyclic nucleotide levels induced by functional synaptic transmission or by longer term behavioral or environmental effects within the brain can result in measurable enzyme induction (see KAUFFMAN et al., 1972). However, glioma cells exposed to norepinephrine exhibit induction of a new phosphodiesterase isozyme (UZUNOV et al., 1973), an effect which may underlie the recent observations by DEVELLIS and BROOKER (1974) that inhibitors of protein or RNA synthesis will prevent refractoriness in the ability of norepinephrine to increase cyclic AMP levels in these cells.

Several other reports have indicated that cultured neuroblastoma cells exposed to cyclic AMP also exhibit increased levels of the enzyme tyrosine hydroxylase, the enzyme step assumed to be rate limiting (see MOLINOFF and AXELROD, 1971; AXELROD, 1971) in the synthesis of catecholamines (WAYMIRE et al., 1972; RICHELSON, 1973). Intracerebral cyclic AMP may also influence the analogous step in serotonin synthesis (TAGLIAMONTE et al., 1971).

THOENEN, MUELLER and AXELROD (MUELLER et al., 1969; THOENEN et al., 1969; THOENEN, 1970; THOENEN et al., 1974) had demonstrated earlier that a delayed induction of tyrosine hydroxylase occurs following increased synaptic activity in the adrenal medulla or superior cervical sympathetic ganglion. This phenomenon with tyrosine hydroxylase can be induced by such whole animal perturbations as prolonged cold exposure, immobilization stress, or the sympathetic hypo-function which follows the administration of reserpine, 6-hydroxydopamine or aminophylline. Moreover, the increase in tyrosine hydroxylase activity represents a true inductive synthesis (see AXELROD, 1971; CHUANG and COSTA, 1974).

Although none of these events were initially related to possible cyclic nucleotide-influenced changes, GUIDOTTI and COSTA (1973) (GUIDOTTI et al., 1973) subsequently reported that the same sets of physical or pharmacological manipulations which lead to the delayed induction of adrenal medullary tyrosine hydroxylase, also promptly elevate the cyclic AMP content of this tissue. GUIDOTTI and COSTA (1973) suggested that the induction of the enzyme by increased activity of cholinergic nicotinic synapses on adrenal medullary cells could involve the activation of one or more protein phosphotransferase reactions by the elevation of cyclic AMP (see COSTA et al., 1974). As a result of this suggestion, several additional lines of experimentation have been reported, in which the role of cyclic AMP in the trans-synaptic induction of tyrosine hydroxylase in both sympathetic ganglia and adrenal medulla has been studied (COSTA et al., 1974; GUIDOTTI et al., 1973, 1975; OTTEN et al., 1973, 1974; MUELLER et al., 1974; THOENEN et al., 1974).

Although no consensus has yet been reached in the interpretation of these manipulations in vivo, several crucial features of the stress and of the sympathetic hypotonia which can induce the enzyme induction have been demonstrated. Thus, it would appear that cortico-steroids released during the stress may be able to lead to an induction of ganglionic tyrosine hydroxylase whether post-ganglionic neuronal cyclic AMP rises or not (COSTA et al., 1974; KEEN and MCLEAN, 1974; see also OTTEN et al., 1974). However, sympathetic ganglia cultured under conditions of stabilized neuronal populations do exhibit increased tyrosine hydroxylase activity upon exposure to exogenous cyclic AMP (MCKAY and IVERSEN, 1972). Furthermore, in the adrenal medulla, delayed induction of tyrosine hydroxylase seems closely related to changes in cyclic AMP content when these changes last for a sufficiently long period to result in the activation of protein kinase activity (GUIDOTTI et al., 1975). Reserpine treatment has also been reported to increase the tyrosine hydroxylase within the central norepinephrine containing neurons of the locus coeruleus after 48–96 hours (ZIGMOND et al., 1974), but no relation to synaptic effects was proposed.

In a shorter term example of enzyme regulation, GOLDSTEIN (1973) and his colleagues (ANAGNOSTE et al., 1974) observed that cyclic AMP would enhance acutely the dopamine synthesis of caudate nucleus slices. This effect seems unrelated to the receptors by which dopamine can activate cyclic AMP synthesis in the same brain region, since neither dopamine agonists or antipsychotic butyrophenones will interact with the effects of cyclic AMP on dopamine synthesis. Subsequently, HARRIS et al. (1974) reported that cyclic AMP and 4 derivatives of cyclic AMP with either increased resistance to phosphodiesterase or increased activity on protein kinase, would rapidly (20–30 min) increase the tyrosine hydroxylase activity of rat caudate nucleus homogenate fractions enriched in synaptosomes (see also GOLDSTEIN et al., 1973). Since these effects of exogeneous cyclic AMP occur quite rapidly, enzyme induction seems unlikely. These changes could represent the acute results of a phosphotransferase reaction in which the substrate might be either the hydroxylase enzyme itself or one of the enzymes which produces the co-factors necessary for hydroxylation to occur. Based on the rapid effects of nerve stimulation in activating the tyrosine hydroxylase of the guinea pig vas deferens, Roth and associates (MORGENROTH et al., 1974) have suggested that the activity-related influx of Ca^{++} mediates the activation. In relation to factors discussed above (III.D.3), this influx could regulate or dissociate proteins which bind calcium from the tyrosine hydroxylase or a related protein which would regulate the enzyme or a co-factor.

Inasmuch as dopamine and dopamine agonists do not interact directly with cyclic AMP on the stimulation of adenylate cyclase, an effect upon some cofactor generating system may be more likely (see EBSTEIN et al., 1974). However, in keeping with other conceptual models for the short term regulation of tyrosine hydroxylase by such events as end product inhibition (see MOLINOFF and AXELROD, 1971), it will be important to learn whether the effect of cyclic AMP on the tyrosine hydroxylating activity (presumably a presynaptic event) bears any relationship to either the alpha or beta catecholamine-sensitive adenylate cyclases which can only be observed in rat caudate slices (FORN et al., 1974; but see WALKER and WALKER, 1973a, b) presumed to represent post-synaptic responses.

The continuous activity along this exciting line of highly competitive research suggests that much more will soon be learned about these interacting regulatory relationships. It is of interest in this regard that the pineal gland exhibits markedly increased synthesis of the secretory product melatonin after incubation in vitro with either norepinephrine or cyclic AMP (KLEIN et al., 1973) and that norepinephrine will activate the adenylate cyclase of the pineal via a beta receptor (see WEISS and COSTA, 1968; WEISS and KIDMAN, 1969). However, because the increases in apparent melatonin synthesis caused by added norepinephrine or cyclic AMP were accompanied by decreased pineal content of the precursor serotonin (but also see SHEIN and WURTMAN, 1969), it seems likely that the cyclic AMP-induced activation occurs at the step of serotonin-N-acetyltransferase (KLEIN et al., 1973; also see DEGUCHI and AXELROD, 1973), rather than at that of tryptophane hydroxylase (but see SHEIN and WURTMAN, 1969).

VII. Conclusions

Despite massive efforts, the role and mechanism by which cyclic nucleotides are involved in synaptic function and other longer term neurobiological phenomena can now only begin to be sketched into perspective. The evidence associating cyclic nucleotides with the molecular process of synaptic transmission is presently concentrated from three sets of central synaptic junctions and one class of peripheral sympathetic ganglia. In the CNS, biochemical, cytochemical, and electrophysiological experiments indicate that the norepinephrine neurons of the nucleus locus coeruleus inhibit Purkinje cells of the cerebellar cortex and pyramidal cells of the hippocampus in a manner identical to the inhibition produced by exogenous cyclic AMP: the target neurons are hyperpolarized with increased membrane resistance. In these two target areas and in the dopamine-containing inhibitory pathway between substantia nigra and caudate nucleus, the endogenous catecholamine is known to activate adenylate cyclase activity and elevate cyclic AMP content. An increased content of cyclic AMP in cerebellar Purkinje cells during the action of the synaptic pathway can also be demonstrated by immunocytochemical studies. Moreover, the effects of the catecholamines are potentiated by phosphodiesterase inhibitors, and antagonized by substances which—at least in the peripheral nervous system—antagonize the effects of the catecholamine on cyclic AMP synthesis.

In sympathetic ganglia, dopamine can also regulate cyclic AMP synthesis while acetylcholine has been shown to regulate cyclic GMP synthesis; furthermore, each cyclic nucleotide can produce specific biophysical responses similar to slow inhibitory and excitatory synaptic potentials, respectively. The dopamine-induced hyperpolarizing response is blocked by the same drugs which prevent the activation of cyclic AMP synthesis and is potentiated by phosphodiesterase inhibition. In the sympathetic ganglion, and in the cerebral cortex, there is also preliminary evidence that cholinergic excitatory transmission can be emulated by cyclic GMP, but no extended pharmacological or electrophysiological analysis of this effect has yet been reported.

Finally, the effect of the cyclic AMP or cyclic GMP produced at these sites of synaptic transmission is now considered to be mediated through additional intracellular sequences, involving protein kinase and probably Ca^{++} as well. Preliminary experiments show that the physiological potency of exogenous synthetic cyclic AMP derivatives varies in close correlation with their ability to activate cyclic AMP-dependent protein kinases. Moreover, the iontophoresis of cyclic AMP intracellularly can produce the same pattern of biophysical responses in hippocampal pyramidal cells as seen when norepinephrine is applied extracellularly or when the norepinephrine pathway is activated.

A. Information Processing and the Functional Regulation of Neuronal Metabolism

The range of functions which could be controlled through cyclic nucleotide-sensitive protein kinases may currently be said to know no limits, since so few studies have as yet been reported in which substrate proteins of known function have been evaluated. Available evidence strongly suggests that some such effects can occur on uncharacterized synaptic membrane proteins, on transmitter-related synthetic enzymes, and possibly on other proteins as well. In the face of such a powerful biochemical lever for the regulation of neuronal and possibly glial metabolism, it might be wondered whether the electrophysiological effects of postsynaptic inhibition represent the primary message of such synaptic events or whether these electrophysiological effects might not be epi-phenomena of a more pervasive, but covert, shift in cellular metabolism which is evoked by these cyclic nucleotide-mediated synaptic stimuli.

The same general question can be approached from the perspective of cells in other tissues where hormones produce events in which electrical signs are never viewed as the major end product. For example, hyperpolarization also occurs when adrenal cortical cells are activated by ACTH (see RUBIN et al., 1972; SAYERS et al., 1972; MATTHEWS and SAFFRAN, 1973; DAMBACH and FRIEDMANN, 1974; and V.D) or when hepatocytes are stimulated with glucagon or pyruvate, but these cells are thought to be electrically "inexcitable". Thus, even though the electrical activity of neurons has long been their most striking functionally-related property, electrical changes which accompany synaptic events may not be the primary objective of every synaptic transmission. Some further speculation along this line may help to sharpen the image of "supra-electrical" events in which cyclic nucleotides may be involved.

Synapses which produce a postsynaptic electrophysiological change mediated by cyclic nucleotides may have been important during differentiation and persisted into adulthood as vestiges to yield intermittent low level trophic signals (see MAEDA et al., 1974), which are generated only after multiple repetitions such as the intense firing of some norepinephrine neurons during certain stages of sleep (JOUVET, 1972; CHU and BLOOM, 1973, 1974b). It may be pertinent to this trophic viewpoint, that the locus coeruleus neurons retain into adulthood the capacity to form collateral axon sprouts (BJÖRKLUND et al., 1971; KATZMAN et al., 1971; MOORE et al., 1971; BLOOM et al., 1974b). Alternatively, the cyclic nucleotide-intracellular sequence may represent a form of intercellular chemical communication which evolved early in phylogeny and which now serves to amplify the effects of relatively small populations of neurons, like the catecholamine neurons, which extend diffusely throughout the nervous system. This chemical amplification could compensate for their limited number (caused perhaps by their schedule of differentiation) in relation to the large number of target cells they must serve. In yet a third scheme, neurons whose synaptic effects are mediated through cyclic nucleotide amplifying steps may do so to "bias" more rapidly transmitting systems which project to the same target neurons, and in which the resultant changes in membrane and intracellular proteins could accentuate or suppress previously established input and output relationships. Under this view, synaptic events which result in intracellular changes in cyclic AMP or cyclic GMP could be considered holistically to transform the postsynaptic cell from one level of metabolic functioning to another by altering one or more interactive enzyme-substrate relationships. For example, in the heart, catecholamines not only increase the force and frequency of the cardiac contractions, they also active lipolysis and glycogenolysis in order to provide the heart with increased substrate levels for energy metabolism (see MAYER, 1974).

Most attempts to evaluate the possible influence of cyclic AMP on the activation of brain glycogenolysis have instead indicated that conditions under which cyclic AMP levels are elevated do not increase the already rapid rates of glycogenolysis nor is the relative amount of phosphorylase increased (GOLDBERG and O'TOOLE, 1969; KAKIUCHI and RALL, 1968a; KNULL et al., 1974; but also see FOLBERGROVA, 1974). This should not be unexpected since brain stores of glycogen are relatively evanescent and experimental manipulation of these levels without rapid freeze or other physical fixations may not be able to reveal the control mechanisms in vivo. Such an activation of glycogen breakdown could accompany neuronal responses to a transmitter like norepinephrine, even though the direct effects of the transmitter on cell firing are depressant. Thus, the "depressant" effect of catecholamines in the central nervous system may represent a holistic mechanism by which one minor population of synaptic inputs can tune the metabolic machinery of the postsynaptic neuron for the modulation or requirements of other types of synaptic communication. Such an effect of overt inhibition might then be taken as a possible sign of a more complex chemical vocabulary by which nerve cells may communicate (BLOOM, 1973b).

B. Cyclic Nucleotides: Molecular Regulators in a Multi-Determinant Matrix

The intensive obsession of this review with the cyclic nucleotides should not obscure recognition of the fact that these substances occur in cells at less than $^{1}/_{1000}$ the concentration of ATP and at perhaps an even smaller fractional concentration of the free intracellular Ca^{++}. In noting this marked concentration difference between cyclic nucleotides and other nucleotides, SUTHERLAND (see ROBISON et al., 1971) observed that ATP, and perhaps also Ca^{++}, was absolutely essential for every cell function, in the sense that the mechanisms for maintaining cellular viability depended upon the adequate production of ATP and the rapid translocations of Ca^{++}. Nevertheless, cyclic nucleotides working within cells at relatively low concentrations have also been documented to influence specific and discrete aspects of cell function. In order to do so, they must, therefore, interact with all of the other modulatory chemicals which can impinge upon the enzymes and substrates by which cyclic nucleotides themselves are regulated and by which their effects arise (Fig. 29). The complex interrelationships schematized in this diagram suggest only the minimum of modes by which cyclic nucleotides, Ca^{++} and ATP interact to regulate almost every aspect of neuronal function from membrane properties to genetic expression. By substituting a soluble guanylate cyclase (perhaps loosely associated with the plasma membrane) a similar interreactive network could apply to the functional roles mediated by cyclic GMP.

One overly simplified abstract conceptualization as to how cyclic nucleotides operate on such events is a two dimensional matrix, in which cellular events are grouped in terms of their independence of or sensitivity to any of the presently known regulators (either cyclic AMP or cyclic GMP versus Ca^{++}); for example, a protein kinase responsible for phosphorylation of a membrane protein might be cyclic AMP dependent and Ca^{++}-resistant or cyclic GMP-dependent and Ca^{++}-inhibited. Furthermore, interactions between any two such regulating processes must also interact with infinite additional dimensions of known regulating molecules and ions and others still undiscovered. Earlier sections have emphasized the powerful sequential relationships and disassociations. Although purely abstract at present, such a multi-dimensional matrix of interactive regulating systems highlights the organic nature of the neuron as a living cell and not a transistor, and also emphasizes the large number of molecular specifications by which neuronal events could be coded. Furthermore, comprehensive integrated views of the quantitative relationship between energy producing and energy consuming metabolic steps suggest that—in liver cells—a series of interactive equilib-

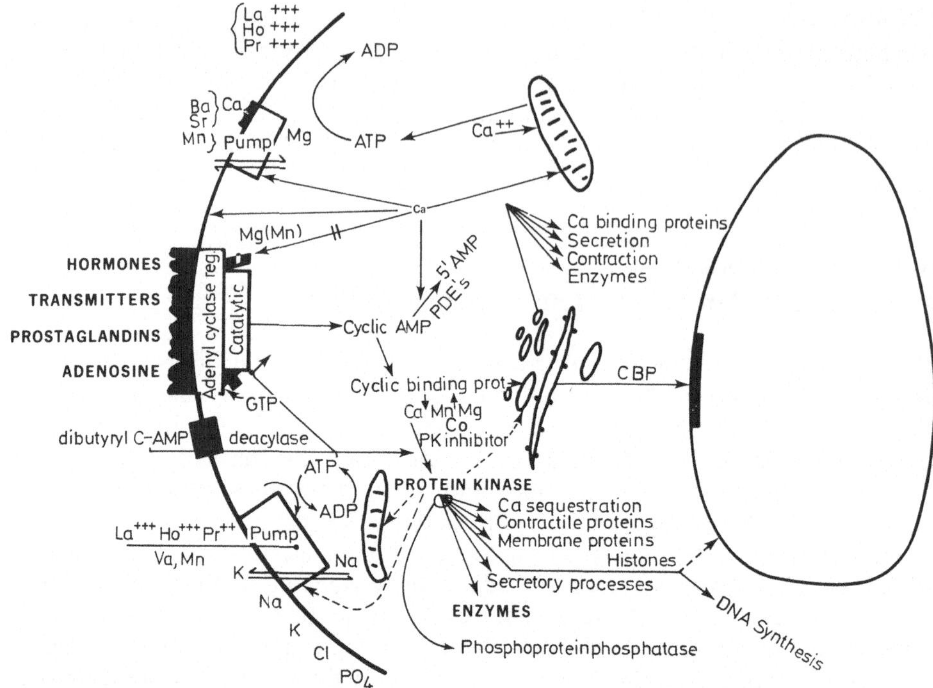

Fig. 29. Schematic representation of the multiple lines of inter-reactive systems by which cyclic nucleotides, ATP, and metallic cations can influence the functional status of biochemical and bioelectrical properties of neurons. The multivalent cations used experimentally to interfere with Ca^{++} fluxes and Ca^{++} dependent monovalent pumps are indicated for future references, although such data for central neurons is still lacking. Dibutyryl cyclic AMP is shown entering the cell after partial de-acylation, but this de-acylase is relatively inactive in brain. Ca^{++} related events (storage, binding and translocation) are emphasized in the upper half of the hypothetical cell, while cyclic AMP related events are emphasized in the lower half. The inter-reactions on adenylate cyclase substrates, co-factors and protein kinases are indicated at the interface between the two systems. For further details see text

ria between shared products and substrates results in a dynamic kinetic functional state in which impressed hormonal signals may be able to elicit needed changes in ongoing reaction rates quite rapidly (see ATKINSON, 1968; WILSON et al., 1974). Similar rapid events would seem even more critical to the needs of the neuron, but whether cyclic nucleotides represent an early or a late development in the evolution of this interactive regulatory-response matrix must await still further research.

VIII. Acknowledgements

I thank my colleagues, Drs. GEORGE SIGGINS, BARRY HOFFER, and JAMES NATHANSON, for many helpful discussions and Dr. PAUL GREENGARD for his meticulous reading of the manuscript. I thank Ms. ELENA BATTENBERG for her unflagging efforts with the bibliography, and Mrs. ODESSA COLVIN for her tireless typing. I thank Drs. GREENGARD, ADINOLPHI, SIGGINS, HOFFER, FREEDMAN and SEGAL, for permission to reprint illustrations of their work. Lastly, I thank the publishers of Brain Research, Science, the Journal of Neurobiology, the Journal of Pharmacology and Experimental Therapeutics and the Journal of Cyclic Nucleotide Research for permission to republish our figures.

References

ABDULLA, Y.H., HAMADAH, K.: 3',5'-Cyclic adenosine monophosphate in depression and mania. Lancet **1970 I**, 378–381.

ADINOLFI, A.M., SCHMIDT, S.Y.: Cytochemical localization of cyclic nucleotide phosphodiesterase activity at developing synapses. Brain Res. **76**, 21–31 (1974).

AGHAJANIAN, G.K., HAIGLER, H.J., BLOOM, F.E.: Lysergic acid diethylamide and serotonin: direct actions on serotonin-containing neurons in rat brain. Life Sci. **11**, 615–622 (1972).

AMER, M.S., KREIGHBAUM, E.: Cyclic nucleotide phosphodiesterase: properties, activators, inhibitors, structure-activity relationships and possible role in drug development. J. pharm. Sci. **64**, 1–37 (1975).

ANAGNOSTE, B., SHIRRON, C., FRIEDMAN, E., GOLDSTEIN, M.: Effect of dibutyryl cyclic adenosine monophosphate on C^{14} dopamine synthesis in rat brain striatal slices. J. Pharmacol. exp. Ther. **191**, 370–376 (1974).

ANDERSON, E.G., HAAS, H., HOSLI, L.: Comparison of effects of noradrenaline and histamine with cyclic AMP on brain stem neurones. Brain Res. **49**, 471–475 (1973a).

ANDERSON, W.B., RUSSELL, T.R., CARCHMAN, R.A., PASTAN, I.: Interrelationship between adenylate cyclase activity, adenosine 3',5'-cyclic monophosphate levels, and growth of cells in culture. Proc. nat. Acad. Sci. (Wash.) **70**, 3802–3805 (1973b).

APPLEMAN, M.M., TERASAKI, W.L.: The regulation of cyclic nucleotide phosphodiesterase. In: Advances in Cyclic Nucleotide Research, vol. 5. New York: Raven Press (in press).

APPLEMAN, M.M., THOMPSON, W.J., RUSSELL, T.R.: Cyclic nucleotide phosphodiesterases. In: Advances in Cyclic Nucleotide Research, vol. 3. New York: Raven Press 1973.

ARBUTHNOTT, G.W., ATTREE, T.J., ECCLESTON, D., LOOSE, R.W., MARTIN, M.J.: Is adenylate cyclase the dopamine receptor. Med. biol. Ill. **52**, 350–353 (1974).

ASCHER, P.: Excitatory effects of dopamine on molluscan neurons. In: Frontiers in Catecholamine Research. New York: Pergamon Press 1973.

ASHBY, C.D., WALSH, D.A.: Characterization of the interaction of a protein inhibitor with adenosine 3',5'-monophosphate-dependent protein kinases. J. biol. Chem. **247**, 6637–6642 (1972).

ASHMAN, D.F., LIPTON, R., MELICOW, M.M., PRICE, T.D.: Isolation of adenosine 3',5'-monophosphate and guanosine 3',5'-monophosphate from rat urine. Biochem. biophys. Res. Commun. **11**, 330–334 (1963).

ATKINSON, D.E.: Citrate in the regulation of energy metabolism. In: Metabolic Roles of Citrate. New York: Academic Press 1968.

AURBACH, G.D., FEDAK, S.A., WOODARD, C.J., PALMER, J.S., HAUSER, D., TROXLER, F.: β-adrenergic receptor: stereospecific interaction of iodinated β-blocking agent with high affinity site. Science **186**, 1223–1224 (1974).

AXELROD, J.: Noradrenaline: fate and control of its biosynthesis. Science **173**, 598–606 (1971).

BARKER, J.L., CRAYTON, J., NICOLL, R.: Noradrenaline and acetylcholine responses of supraoptic neurosecretory cells. J. Physiol. (Lond.) **218**, 19–32 (1971).

BATZRI, S., SELINGER, Z., SCHRAMM, M.: Potassium ion release and enzyme secretion: adrenergic regulation by α and β receptors. Science **174**, 1029–1031 (1971).

BAUDRY, M., MARTRES, M-P., SCHWARTZ, J-C.: H_1 and H_2 receptors in the histamine-induced accumulation of cyclic AMP in guinea pig slices. Nature (Lond.) **253**, 362–363 (1975).

BAYLOR, D.A., FUORTES, M.G.F.: Electrical responses of single cones in the retina of the turtle. J. Physiol. (Lond.) **207**, 77–92 (1970).

BEAVO, J.A., HARDMAN, J.G., SUTHERLAND, E.W.: Hydrolysis of cyclic guanosine and adenosine 3′,5′-monophosphate by rat and bovine tissues. J. biol. Chem. **245**, 5649–5655 (1970).

BEER, B., CHASIN, M., CLODY, D.E., VOGEL, J.R., HOROVITZ, Z.P.: Cyclic adenosine monophosphate phosphodiesterase in brain: Effect on anxiety. Science **176**, 428–430 (1972).

BENSINGER, R.E., FLETCHER, R.T., CHADER, G.J.: Guanylate cyclase: inhibition by light in retinal photoreceptors. Science **183**, 86–87 (1974).

BERG, J.S.V.: Inhibitory effects of dibutyryl and cyclic AMP on the compound action potential in the frog (Rana Pipiens) sciatic nerve. Experientia (Basel) **30**, 1025–1026 (1974).

BERGER, B., TASSIN, J.P., BLANC, O., MOYNE, M.A., THIERRY, A.M.: Histochemical confirmation for dopaminergic innervation of rat cerebral cortex after destruction of the noradrenergic ascending pathways. Brain Res. **81**, 332–337 (1974).

BERKOWITZ, B.A., TARVER, J.H., SPECTOR, S.: Release of norepinephrine in the central nervous system by theophylline and caffeine. Europ. J. Pharmacol. **10**, 64–71 (1970).

BERNE, R.M., RUBIO, R., CURNISH, R.R.: Release of adenosine from ischemic brain. Circulat. Res. **35**, 262–271 (1974).

BERRIDGE, M.J., PRINCE, W.T.: The electrical response of isolated salivary glands during stimulation with 5-hydroxytryptamine and cyclic AMP. Phil. Trans. B **262**, 111–120 (1971).

BERRIDGE, M.J., PRINCE, W.T.: Transepithelial potential changes during stimulation of isolated salivary glands with 5-hydroxytryptamine and cyclic AMP. J. exp. Biol. **56**, 139–153 (1972).

BERTI, F., TRABUCCHI, M., BERNAREGGI, V., FUMAGALLI, R.: The effect of prostaglandins on cyclic AMP formation in cerebral cortex of different mammalian species. Pharmacol. Res. Commun. **4**, 253–259 (1972).

BIANCHI, C.P.: Cell Calcium. New York: Appleton-Century-Crofts 1964.

BILZEKIAN, J.P., AURBACH, G.D.: The effects of nucleotides on the expression of β-adrenergic adenylate cyclase activity in membranes from turkey erythrocytes. J. biol. Chem. **249**, 157–161 (1974).

BIRNBAUMER, L., YANG, P.C.: Studies on receptor-mediated activation of adenylyl cyclases. I. Preparation and description of general properties of an adenylyl cyclase system in beef renal medullary membranes sensitive to neurohypophyseal hormones. J. biol. Chem. **249**, 7848–7856 (1974a).

BIRNBAUMER, L., YANG, P.C.: Studies on receptor-mediated activation of adenylyl cyclases. III. Regulation by purine nucleotides of the activation of adenylyl cyclases from target organs for prostaglandins, luteinizing hormone, neurohypophyseal hormones and catecholamines. Tissue and hormone-dependent variations. J. biol. Chem. **249**, 7867–7873 (1974b).

BIRNBAUMER, L., NAKAHARA, T., YANG, P.C.: Studies on receptor-mediated activation of adenylyl cyclases. II. Nucleotide and nucleoside regulation of the activities of the beef renal medullary adenylyl cyclase and their stimulation by neurohypophyseal hormones. J. biol. Chem. **249**, 7857–7866 (1974).

BITENSKY, M.W., GORMAN, R.E., MILLER, W.H.: Adenyl cyclase as a link between photon capture and changes in membrane permeability of frog photoreceptors. Proc. nat. Acad. Sci. (Wash.) **68**, 561–562 (1971).

BITENSKY, M.W., MIKI, N., MARCUS, F.R., KEIRNS, J.J.: The role of cyclic nucleotides in visual excitation. Life Sci. **13**, 1451–1472 (1973).

BJÖRKLUND, A., CEGRELL, L., FALCK, B., RITZEN, M., ROSENGREN, E.: Dopamine-containing cells in sympathetic ganglia. Acta physiol. scand. **78**, 334–338 (1970).

BJÖRKLUND, A., KATZMAN, R., STENEVI, U., WEST, K.: Development and growth of axonal sprouts from NA and 5-hydroxytryptamine neurons in rat spinal cord. Brain Res. **31**, 21–33 (1971).

BLACK, A.C., BHALLA, R.C., WILLIAMS, T.H.: Species differences in the adenyl cyclase responsiveness to neurotransmitters in the superior cervical ganglion. Abstr. 4th Annu. Meeting Soc. Neurosci., St. Louis, p. 144 (1974).

BLECHER, M., HUNT, N.H.: Enzymatic deacylation of mono- and dibutyryl derivatives of cyclic adenosine 3′,5′-monophosphate by extracts of rat tissues. J. biol. Chem. **247**, 7479–7484 (1972).

BLOOM, F.E.: Electrophysiological pharmacology of single nerve cells. In: Psychopharmacology – A Ten Year Progress Report. Washington, D.C.: U.S. Govt. Printing Office 1968.

BLOOM, F.E.: Fine structural changes in rat brain after intracisternal injection of 6-hydroxydopamine. In: 6-Hydroxydopamine and Catecholamine Neurons. Amsterdam: North Holland Publishing Co. 1971.

BLOOM, F.E.: Amino acids and polypeptides in neuronal function. Neurosci. Res. Program Bull. **10**, 122–251 (1972).

BLOOM, F.E.: Ultrastructural identification of catecholamine-containing central synaptic terminals. J. Histochem. Cytochem. **21**, 333–348 (1973a).

BLOOM, F.E.: Dynamic synaptic communication: finding the vocabulary. Brain Res. **62**, 299–305 (1973b).

BLOOM, F.E.: To spritz or not to spritz: the doubtful value of aimless iontophoresis. Life Sci. **14**, 1819–1834 (1974).

BLOOM, F.E., COSTA, E., SALMOIRAGHI, G.C.: Anesthesia and the responsiveness of individual neurons of the cat's caudate nucleus to acetylcholine, norepinephrine, and dopamine administered by microelectrophoresis. J. Pharmacol. **150**, 244–255 (1965).

BLOOM, F.E., HOFFER, B.J.: Norepinephrine as a central synaptic transmitter. In: Frontiers in Catecholamine Research. New York: Pergamon Press 1973.

BLOOM, F.E., HOFFER, B.J., BATTENBERG, E.F., SIGGINS, G.R., STEINER, A.L., PARKER, C.W., WEDNER, H.J.: Adenosine 3',5'-monophosphate is localized in cerebellar neurons: Immunofluorescence evidence. Science **177**, 436–438 (1972).

BLOOM, F.E., HOFFER, B.J., SIGGINS, G.R.: Studies on norepinephrine containing afferents to Purkinje cells of rat cerebellum. I. Localization of the fibers and their synapses. Brain Res. **25**, 501–521 (1971).

BLOOM, F.E., HOFFER, B.J., SIGGINS, G.R.: Norepinephrine mediated synapses. A model system for neuropsychopharmacology. Biol. Psychiat. **4**, 157–177 (1972).

BLOOM, F.E., KREBS, H., NICHOLSON, J., PICKEL, V.: The noradrenergic innervation of cerebellar Purkinje cells: Localization, function, synaptogenesis, and axonal sprouting of locus coeruleus. In: Dynamics of Degeneration and Growth in Neurons. England: Pergamon Press 1974b.

BLOOM, F.E., SIGGINS, G.R., HOFFER, B.J.: Interpreting the failures to confirm the depression of cerebellar Purkinje cells by cyclic AMP. Science **185**, 627–629 (1974a).

BLOOM, F.E., SIGGINS, G.R., HOFFER, B.J., SEGAL, M., OLIVER, A.P.: The role of cyclic nucleotides in the central synaptic actions of catecholamines. In: Advances in Cyclic Nucleotide Research, vol. 5. New York: Raven Press 1975.

BLOOM, S., SWEAT, F.W.: Covariance of myocardial cyclic AMP and calcium during β-adrenergic stimulation in vivo. Res. Commun. Chem. Pathol. Pharmacol. **8**, 505–514 (1974).

BLUMBERG, J.B., SULSER, F.: The effect of antipsychotic drugs on the cyclic 3',5' adenosine monophosphate system on rat forebrain. Fed. Proc. **33**, 286 (1974).

BOOTH, D.A.: Unlearned and learned effects of intrahypothalamic cyclic AMP injection on feeding. Nature (Lond.) New Biol. **237**, 222–224 (1972).

BORASIO, P.G., VASSALLE, M.: Dibutyryl cyclic AMP and potassium transport in cardiac Purkinje fibers. Amer. J. Physiol. **226**, 1232–1237 (1974).

BORGEAT, P., CHAVANCY, G., DUPONT, A., LABRIE, F., ARIMURA, A., SCHALLY, A.V.: Stimulation of adenosine 3',5'-cyclic monophosphate accumulation in anterior pituitary gland in vitro by synthetic luteinizing hormone-releasing hormone. Proc. nat. Acad. Sci. (Wash.) **69**, 2677–2681 (1972).

BRADSHAW, C.M., SZABADI, E., ROBERTS, M.H.T.: The reflection of ejecting and retaining currents in the time course of neuronal responses to microelectrophoretically applied drugs. J. Pharm. Pharmacol. **25**, 513–520 (1973).

BRAY, J.J., KON, C.M., BRECKENRIDGE, B.M.: Adenyl cyclase cyclic nucleotide phosphodiesterase and axoplasmic flow. Brain Res. **26**, 385–394 (1971).

BRAZEAU, P., VALE, W., BURGUS, R., LING, N., BUTCHER, M., RIUIER, J., GUILLEMIN, R.: Hypothalamic polypeptide that inhibits the secretion of immuno-reactive pituitary growth hormone. Science **179**, 77–79 (1973).

BRECKENRIDGE, B.M.: The measurement of cyclic adenylate in tissues. Proc. nat. Acad. Sci. (Wash.) **57**, 1580–1586 (1964).

BRECKENRIDGE, B.M., BURN, J.H., MATSCHINSKY, F.M.: Theophylline, epinephrine, and neostigmine facilitation of neuromuscular transmission. Proc. nat. Acad. Sci. (Wash.) **57**, 1893–1897 (1967).

BRECKENRIDGE, B.M., JOHNSTON, R.E.: Cyclic 3',5'-nucleotide phosphodiesterase in brain. J. Histochem. Cytochem. **17**, 505–511 (1969).

BRECKENRIDGE, B.M., LISK, R.D.: Cyclic adenylate and hypothalamic regulatory functions. Proc. Soc. exp. Biol. (N.Y.) **131**, 934–935 (1969).

BROOKER, G.: Oscillation of cyclic adenosine monophosphate concentration during the myocardial contraction cycle. Science **182**, 933–934 (1973).

BROSTRÖM, C.O., HUANG, Y-C., BRECKENRIDGE, B. McL., WOLFF, D.J.: Identification of a calcium-binding protein as a calcium-dependent regulation of brain adenylate cyclase. Proc. nat. Acad. Sci. (Wash.) **72**, 64–68 (1975).

BROWN, J.H., MAKMAN, M.H.: Stimulation by dopamine of adenylate cyclase in retinal homogenates and of adenosine-3′,5′-cyclic monophosphate formation in intact retina. Proc. nat. Acad. Sci. (Wash.) **69**, 539–543 (1972).

BROWN, J.H., MAKMAN, M.H.: Influence of neuroleptic drugs and apomorphine on dopamine sensitive adenylate cyclase of retina. J. Neurochem. **21**, 477–479 (1973).

BRUNTON, W.J.: Beta-adrenergic stimulation of transmembrane potential and short circuit current of isolated rabbit oviduct. Nature (Lond.) New Biol. **236**, 12–14 (1972).

BÜLBRING, E., TOMITA, T.: Increase of membrane conductance by adrenaline in the smooth muscle of guinea-pig taenia coli. Proc. roy. Soc. B **172**, 89–102 (1969).

BUNNEY, B.S., AGHAJANIAN, G.K.: Electrophysiological effects of amphetamine in dopaminergic neurons. In: Frontiers in Catecholamine Research. New York: Pergamon Press 1973.

BURGUS, R., GUILLEMIN, R.: Hypothalamic releasing factors. Ann. Rev. Biochem. **39**, 490–526 (1970).

BURNSTOCK, G.: Purinergic nerves. Pharmacol. Rev. **24**, 509–581 (1972).

CARLSOO, B., DANIELSSON, A., MARKLUND, S., STIGBRAND, T.: Effects of 3′,5′-cyclic adenosine monophosphate, 5-hydroxytryptamine, noradrenaline and theophylline on the simultaneous release of peroxidase and amylase from the guinea pig submandibular gland. Acta physiol. scand. **91**, 203–210 (1974).

CARNEGIE, P.R., Kemp, B.E., Dunkley, P.R., Murray, A.W.: Phosphorylation of myelin basic protein by an adenosine 3′:5′-cyclic monophosphate-dependent protein kinase. Biochem. J. **135**, 569–572 (1973).

CARPENTER, D.O., GAUBATZ, G.L.: Octopamine receptors on Aplysia neurones mediate hyperpolarization by increasing membrane conductance. Nature (Lond.) **252**, 483–485 (1974).

CASNELLIE, J.E., GREENGARD, P.: Guanosine 3′:5′-cyclic monophosphate-dependent phosphorylation of endogenous substrate proteins in membranes of mammalian smooth muscle. Proc. nat. Acad. Sci. (Wash.) **71**, 1891–1895 (1974).

CEDAR, H., KANDEL, E.R., SCHWARTZ, J.H.: Cyclic adenosine monophosphate in the nervous system of aplysia californica. I. Increased synthesis in response to synaptic stimulation. J. gen. Physiol. **60**, 558–569 (1972).

CEDAR, H., SCHWARTZ, J.H.: Cyclic adenosine monophosphate in the nervous system of aplysia californica: II. Effect of serotonin and dopamine. J. gen. Physiol. **60**, 570–587 (1972).

CHALAZONITIS, A., GREENE, L.A.: Enhancement in excitability properties of mouse neuroblastoma cells cultured in the presence of dibutyryl cyclic AMP. Brain Res. **72**, 340–345 (1974).

CHASIN, M., MAMRAK, F., SAMANIEGO, S.G.: Preparation and properties of a cell-free, hormonally responsive adenylate cyclase from guinea pig brain. J. Neurochem. **22**, 1031–1038 (1974).

CHASIN, M., MAMRAK, F., SAMANIEGO, S.G., HESS, S.M.: Characteristics of the catecholamine and histamine receptor sites mediating accumulation of cyclic adenosine 3′,5′-monophosphate in guinea pig brain. J. Neurochem. **21**, 1415–1427 (1973).

CHASIN, M., RIVKIN, I., MAMRAK, F., SAMANIEGO, G., HESS, S.M.: α- and β-adrenergic receptors as mediators of accumulation of cyclic adenosine 3′,5′-monophosphate in specific areas of guinea pig brain. J. biol. Chem. **246**, 3037–3041 (1971).

CHATZKEL, S., ZIMMERMAN, I., BERG, A.: Modulation of cyclic AMP synthesis in the cat superior cervical ganglion by short term presynaptic stimulation. Brain Res. **80**, 523–526 (1974).

CHEUNG, W.Y.: Properties of cyclic 3′,5′-nucleotide phosphodiesterase from rat brain. Biochemistry (Wash.) **6**, 1079–1087 (1970).

CHIARANDINI, D.J., BENTLEY, P.J.: The effects of verapamil on metabolism and contractility of the toad skeletal muscle. J. Pharmacol. exp. Ther. **186**, 52–59 (1973).

CHOU, W.S., HO, A.K.S., LOH, H.H.: Neurohormones on brain adenyl cyclase activity in vivo. Nature (Lond.) New Biol. **233**, 280–281 (1971).

CHU, N-S., BLOOM, F.E.: Norepinephrine-containing neurons: changes in spontaneous discharge patterns during unrestrained sleeping and waking. Science **179**, 908–910 (1973).

CHU, N-S., BLOOM, F.E.: The catecholamine-containing neurons in the cat dorso-lateral pontine tegmentum: distribution of the cell bodies and some axonal projections. Brain Res. **66**, 1–21 (1974a).

CHU, N-S., BLOOM, F.E.: Activity patterns of catecholamine-containing pontine neurons in the dorsolateral tegmentum of unrestrained cats. J. Neurobiol. **5**, 527–544 (1974b).

CHUANG, D-M., COSTA, E.: Biosynthesis of tyrosine hydroxylase in rat adrenal medulla after exposure to cold. Proc. nat. Acad. Sci. (Wash.) **71**, 4570–4574 (1974).

CLARK, R.B., GROSS, R., SU, Y-F., PERKINS, J.P.: Regulation of adenosine 3′,5′-monophosphate content in human astrocytoma cells by adenosine and adenine nucleotides. J. biol. Chem. **249**, 5296–5303 (1974).

CLARK, R.B., PERKINS, J.P.: Regulation of adenosine 3′,5′-monophosphate concentration in cultured human astrocytoma cells by catecholamines and histamine. Proc. nat. Acad. Sci. (Wash.) **68**, 2757–2760 (1971).

CLARK, W.G., CUMBY, H.R., DAVIS, H.E.: The hyperthermic effect of intracerebroventricular cholera enterotoxin in the unanesthetized cat. J. Physiol. (Lond.) **240**, 493–504 (1974).

CLARKE, G., HILL, R.G., SIMMONDS, M.A.: Microiontophoretic release of drugs from micropipettes: use of ^{24}Na as a model. Brit. J. Pharmacol. **48**, 156–161 (1973).

CLEMENT-CORMIER, Y.C., KEBABIAN, J.W., PETZOLD, G.I., GREENGARD, P.: Dopamine-sensitive adenylate cyclase in mammalian brain: a possible site of action of antipsychotic drugs. Proc. nat. Acad. Sci. (Wash.) **71**, 1113–1171 (1974).

COHN, M.L., COHN, M., TAYLOR, F.H.: Norepinephrine – an antagonist of dibutyryl cyclic AMP in the regulation of narcosis in the rat. Res. Commun. Chem. Pathol. Pharmacol. **7**, 687–699 (1974).

CONNOR, J.D.: Caudate nucleus neurones: correlation of the effects of substantia nigra stimulation with iontophoretic dopamine. J. Physiol. (Lond.) **208**, 691–703 (1970).

CONTRERAS, E., CASTILLO, S., QUIJADA, L.: Effect of drugs that modify 3′-5′-AMP concentrations on morphine analgesia. J. Pharm. Pharmacol. **24**, 65–66 (1972).

CORRODI, H., FUXE, K., JONSSON, G.: Effects of caffeine on central monoamine neurons. J. Pharm. Pharmacol. **24**, 155–158 (1972).

COSTA, E., GUIDOTTI, A., HANBAUER, I.: Do cyclic nucleotides promote the trans-synaptic induction of tyrosine hydroxylase. Life Sci. **14**, 1169–1188 (1974).

COUTEAUX, R.: Localization of cholinesterases at neuromuscular junctions. Int. Rev. Cytol. **4**, 335–375 (1955).

CRAIN, S.M., POLLACK, E.D.: Restorative effects of cyclic AMP on complex bioelectric activities of cultured fetal rodent CNS tissues after acute Ca^{++} deprivation. J. Neurobiol. **4**, 321–342 (1973).

CRAMER, H., JOHNSON, D.G., HANBAUER, I., SILBERSTEIN, S.D., KOPIN, I.J.: Accumulation of adenosine 3′,5′-monophosphate induced by catecholamines in the rat superior cervical ganglion In Vitro. Brain Res. **53**, 97–104 (1973).

CRAMER, H., NG, L.K.Y., CHASE, T.N.: Effect of probenecid on levels of cyclic AMP in human cerebro-spinal fluid. J. Neurochem. **19**, 1601–1602 (1972).

CUATRECASAS, P.: Interaction of vibrio cholerae enterotoxin with cell membranes. Biochemistry **12**, 3547–3581 (1973).

CUATRECASAS, P.: Membrane receptors. Ann. Rev. Biochem. **45**, 169–214 (1974).

CUATRECASAS, P.: Hormone receptors – their function in cell membranes and some problems related to methodology. In: Advances in Cyclic Nucleotide Research, vol. 5. New York: Raven Press 1975.

CURTIS, D.R.: Microelectrophoresis. In: Physical Techniques in Biological Research, vol. 5. New York: Academic Press 1964.

CURTIS, D.R., JOHNSTON, G.A.R.: Amino acid transmitters in the mammalian central nervous system. Ergebn. Physiol. **69**, 97–188 (1974).

DALE, H.H., FELDBERG, W., VOGT, M.: Release of acetylcholine at voluntary motor nerve endings. J. Physiol. (Lond.) **86**, 353–380 (1936).

DALTON, C., CROWLEY, H.J., SHEPPARD, H., SCHALLEK, W.: Regional cyclic nucleotide phosphodiesterase activity in cat central nervous system: effects of benzodiazepines. Proc. Soc. exp. Biol. (N.Y.) **145**, 407–410 (1974).

DALY, J.: The role of cyclic nucleotides in the nervous system. In: Handbook of Psychopharmacology. New York: Plenum Press 1975.

DALY, J.W., HUANG, M., SHIMIZU, H.: Regulation of cyclic AMP levels in brain tissue. In: Advances in Cyclic Nucleotide Research, vol. 1. New York: Raven Press 1972.

DAMBACH, G., FRIEDMANN, N.: Substrate-induced membrane potential changes in the perfused rat liver. Biochim. biophys. Acta (Amst.) **367**, 366–370 (1974).

DAVIDOFF, R.A.: Gamma amino butyric acid antagonism and presynaptic inhibition in the frog spinal cord. Science **175**, 331–333 (1972).

DEGUCHI, T., AXELROD, J.: Superinduction of serotonin N-acetyl-transferase and supersensitivity of adenyl cyclase to catecholamines in denervated pineal gland. Molec. Pharmacol. **9**, 612–618 (1973).

DELCASTILLO, J., KATZ, B.: A comparison of acetylcholine and stable depolarizing agents. Proc. roy. Soc. B **146**, 362–368 (1957).

DHALLA, N.S., SULAKHE, P.V., MCNAMARA, D.B.: Studies on the relationship between adenylate cyclase activity and calcium transport by cardiac sarcotubular membranes. Biochim. biophys. Acta (Amst.) **323**, 276–284 (1973).

DIAMOND, J.: Phosphorylase, calcium, and cyclic AMP in smooth-muscle contraction. Amer. J. Physiol. **225**, 930–737 (1973).

DOUSA, T., HECHTER, O.: Lithium and brain adenyl cyclase. Lancet **1970 I**, 834–835.

DRUMMOND, G.I., POWELL, C.A.: Analogues of adenosine 3′,5′-cyclic phosphate as activators of phosphorylase b kinase and as substrates for cyclic 3′,5′-nucleotide phosphodiesterase. Molec. Pharmacol. **6**, 24–30 (1970).

DUN, N., NISHI, S.: Effects of dopamine on the superior cervical ganglion on the rabbit. J. Physiol. (Lond.) **239**, 155–164 (1974).

EBSTEIN, B., ROBERGE, C., TABACHNIK, J., GOLDSTEIN, M.: The effect of dopamine and of apomorphine and dB-cAMP-induced stimulation of synaptosomal tyrosine hydroxylase. J. Pharm. Pharmacol. **26**, 975–977 (1974).

ECCLES, J.C.: The Physiology of Synapses. New York: Academic Press 1964.

ECCLES, R., LIBET, B.: Origin and blockade of the synaptic responses of curarized sympathetic ganglia. J. Physiol. (Lond.) **157**, 484 (1961).

EDSTRÖM, A., KANJE, M., WALUM, E.: Effects of dibutyryl cyclic AMP and prostaglandin E_1 on cultured human glioma cells. Exp. Cell Res. **85**, 217–223 (1974).

EIPPER, B.A.: Rat brain tubulin and protein kinase activity. J. biol. Chem. **249**, 1398–1406 (1974).

ENGBERG, I., FLATMAN, J.A., KADZLELAWA, K.: The hyperpolarisation of motoneurones by electrophoretically applied amines and other agents. Acta physiol. scand. **91**, 3A–4A (1974).

ENGBERG, I., MARSHALL, K.C.: Mechanism of Noradrenaline hyperpolarization in spinal cord motoneurones of the cat. Acta physiol. scand. **83**, 142–144 (1971).

ERANKÖ, O., HARKONEN, M.: Monoamine-containing small cells in the superior cervical ganglion of the rat and one organ composed of them. Acta physiol. scand. **63**, 511–512 (1965).

FALLON, E.F., AGRAWAL, R., FURTH, E., STEINER, A.L., COWDEN, R.: Cyclic guanosine and adenosine 3′,5′-monophosphates in canine thyroid: localization by immunofluorescence. Science **184**, 1089–1091 (1974).

FATT, P., KATZ, B.: An analysis of the endplate potential recorded with an intracellular electrode. J. Physiol. (Lond.) **115**, 320–370 (1951).

FERRENDELLI, J.A., CHANG, M.M., KINSCHERF, D.A.: Elevation of cyclic GMP levels in central nervous system by excitatory and inhibitory amino acids. J. Neurochem. **22**, 535–540 (1974).

FERRENDELLI, J.A., KINSCHERF, D.A., CHANG, M.M.: Regulation of levels of guanosine cyclic 3′,5′-monophosphate in the central nervous system: effects of depolarizing agents. Molec. Pharmacol. **9**, 445–454 (1973).

FERRENDELLI, J.A., KINSCHERF, D.A., KIPNIS, D.M.: Effects of amphetamine, chlorpromazine and reserpine on cyclic GMP and cyclic AMP levels in mouse cerebellum. Biochem. biophys. Res. Commun. **46**, 2114–2120 (1972).

FERRENDELLI, J.A., STEINER, A.L., MCDOUGAL, D.B., KIPNIS, D.M.: The effect of oxotremorine and atropine on cGMP and cAMP levels in mouse cerebral cortex and cerebellum. Biochem. biophys. Res. Commun. **41**, 1061–1067 (1970).

FIELD, M.: Mode of action of cholera toxin: stabilization of catecholamine-sensitive adenylate cyclase in turkey erythrocytes. Proc. nat. Acad. Sci. (Wash.) **71**, 3299–3303 (1974).

FIKUS, M., KWAST-WELFELD, J., KAZI-MIERCZUK, Z., SHUGAR, D.: Biochemical studies on some new analogues of adenosine-3′,5′-cyclic phosphate including isoguanosine-3′5′-cyclic phosphate. Acta biochim. pol. **21**, 465–474 (1974).

FILLER, R., LITWACK, G.: Differences in macromolecular binding between cyclic AMP and its dibutyryl derivative in vitro. Biochem. biophys. Res. Commun. **52**, 159–167 (1973).

FLORENDO, N.T., BARRNETT, R.J., GREENGARD, P.: Cyclic 3′5′-nucleotide phosphodiesterase: cytochemical localization in cerebral cortex. Science **173**, 745–747 (1971).

FOLBERGROVA, J.: Energy metabolism of mouse cerebral cortex during homocysteine convulsions. Brain Res. **81**, 443–454 (1974).

FORN, J., KRISHNA, G.: Effects of biogenic amines on the rate of adenosine 3′,5′ monophosphate formation in brain slices of different animal species. Fed. Proc. **29**, 480 (1970).

FORN, J., KRISHNA, G.: Effect of norepinephrine, histamine, and other drugs on cyclic 3′,5′-AMP formation in brain slices of various animal species. Pharmacology (Basel) **5**, 193–204 (1971).

FORN, J., KRUEGER, B.K., GREENGARD, P.: Adenosine 3′,5′-monophosphate content in rat caudate nucleus. Demonstration of dopaminergic and adrenergic receptors. Science **186**, 1118–1119 (1974).

FORN, J., VALDECASAS, F.G.: Effects of lithium on brain adenyl cyclase activity. Biochem. Pharmacol. **20**, 2773–2779 (1971).

FREDERICKSON, R.C.A., JORDAN, L.M., PHILLIS, J.W.: The action of noradrenaline on cortical neurons: effects of pH. Brain Res. **35**, 556–560 (1971).

FREDERICKSON, R.C.A., JORDAN, L.M., PHILLIS, J.W.: A reappraisal of the actions of noradrenaline and 5-hydroxytryptamine on cerebral cortical neurons. Comp. Gen. Pharmacol. **3**, 443–456 (1972).

FREEDMAN, R., HOFFER, B.J.: Phenothiazine antagonism of the noradrenergic inhibition of cerebellar Purkinje neurons. J. Neurobiol. **6**, 277–288 (1975).

FRIEDMAN, N., SOMLYO, A.V., SOMLYO, A.P.: Cyclic adenosine and guanosine monophosphate and glucagon: effect on liver membrane potentials. Science **171**, 400–402 (1971).

FROEHLICH, J.E., RACHMELER, M.: Effect of adenosine 3′-5′-cyclic monophosphate on cell proliferation. J. Cell. Biol. **55**, 19–31 (1972).

FUMAGALLI, R., BERNAREGGI, V., BERTI, F., TRABUCCHI, M.: Cyclic AMP formation in human brain: an in vitro stimulation by neurotransmitters. Life Sci. **10**, 1111–1115 (1971).

FUXE, K., HÖKFELT, T., JOHANSSON, O., JONSSON, O., LIDBRINK, P., LJUNGDAHL, A.: The origin of dopamine nerve terminals in limbic and frontal cortex. Evidence for meso-cortical dopamine neurons. Brain Res. **82**, 349–355 (1974).

FUXE, K., JONSSON, G.: The histochemical fluorescence method for the demonstration of catecholamines: Theory, practice and application. J. Histochem. Cytochem. **21**, 293–311 (1973).

FUXE, K., OLSON, L., ZOTTERMAN, Y.: Dynamics of Degeneration and Growth in Neurons. Oxford and New York: Pergamon 1974b.

FUXE, K., UNGERSTEDT, U.: Action of caffeine and theophylline on supersensitive dopamine receptors: considerable enhancement of receptor response to treatment with dopa and dopamine agonists. Med. biol. Ill. **52**, 48–54 (1974).

GABBALLAH, S., POPOFF, C.: Cyclic 3′,5′-nucleotide phosphodiesterase in nerve endings of developing rat brain. Brain Res. **25**, 220–222 (1971).

GALINDO, A., KRNJEVIC, K., SCHWARTZ, S.: Microiontophoretic studies on neurones in the cuneate nucleus. J. Physiol. (Lond.) **158**, 296–323 (1967).

GARBARG, M., BARBIN, G., FEGER, J., SCHWARTZ, J-C.: Histaminergic pathway in rat brain evidenced by lesions of the medial forebrain bundle. Science **186**, 833–834 (1974).

GARDNER, D., KANDEL, E.R.: Diphasic postsynaptic potential: a chemical synapse capable of mediating conjoint excitation and inhibition. Science **176**, 675–677 (1972).

GEORGE, W.J., POLSON, J.B., O'TOOLE, A.G., GOLDBERG, N.D.: Elevation of guanosine 3′,5′-cyclic phosphate in rat heart after perfusion with acetylcholine. Proc. nat. Acad. Sci. (Wash.) **66**, 398–403 (1970).

GERGELY, J.: Some aspects of the role of the sarcoplasmic reticulum and the tropomyosin-troponin system in the control of muscle contraction by calcium ions. Circulat. Res. **34** and **35** (Suppl. III) 74–82 (1974).

GIBSON, D.A., REICHLIN, S., VERNADAKIS, A.: ^3H Uridine uptake and incorporation into RNA in the C-6 glial cells following dibutyryl cyclic AMP treatment. Brain Res. **81**, 354–360 (1974).

GILMAN, A.G.: The regulation of cyclic AMP metabolism in cultured cells of the nervous system. In: Advances in Cyclic Nucleotide Research, vol. 1. New York: Raven Press 1972.

GILMAN, A.G., NIRENBERG, M.: Effect of catecholamines on the adenosine 3′,5′-cyclic concentrations of clonal satellite cells of neurons. Proc. nat. Acad. Sci. (Wash.) **68**, 2165–2168 (1971a).

GILMAN, A.G., NIRENBERG, M.: Regulation of adenosine 3′,5′-monophosphate metabolism in cultured neuroblastoma cells. Nature (Lond.) **234**, 356–358 (1971 b).

GILMAN, A.G., SCHRIER, B.K.: Adenosine cyclic 3′,5′-monophosphate in fetal rat brain cell cultures. Molec. Pharmacol. **8**, 410–416 (1972).

GINSBORG, B.L.: Ion movements in junctional transmission. Pharmacol. Rev. **19**, 289–316 (1967).

GODFRAIND, J.M., PUMAIN, R.: Cyclic adenosine monophosphate and norepinephrine: effect on Purkinje cells in rat cerebellar cortex. Science **174**, 1257 (1971).

GOLDBERG, A.L., SINGER, J.J.: Evidence for a role of cyclic AMP in neuromuscular transmission. Proc. nat. Acad. Sci. (Wash.) **64**, 134–141 (1969).

GOLDBERG, N.D.: The Yin Yang hypothesis of biological control: opposing influences of cyclic GMP and cyclic AMP in bidirectionally regulated systems. In: Advances in Cyclic Nucleotide Research, vol. 5. New York: Raven Press (in press).

GOLDBERG, N.D., DIETZ, S.B., O'TOOLE, A.G.: Cyclic guanosine 3′,5′-monophosphate in mammalian tissues and urine. J. biol. Chem. **244**, 4458–4466 (1969).

GOLDBERG, N.D., HADDOX, M.K., HARTLE, D.K., HADDEN, J.W.: The biological role of cyclic 3′,5′-guanosine monophosphate. In: Cellular Mechanisms. Basel: S. Karger 1973 a.

GOLDBERG, N.D., LUST, W.D., O'DEA, R.F., WEI, S., O'TOOLE, A.G.: A role of cyclic nucleotides in brain metabolism. Advanc. Biochem. Psychopharmacol. **3**, 67–87 (1970).

GOLDBERG, N.D., O'DEA, R.F., HADDOX, M.K.: Cyclic GMP. In: Advances in Cyclic Nucleotide Research, vol. 3. New York: Raven Press 1973 b.

GOLDBERG, N.D., O'TOOLE, A.G.: The properties of glycogen synthetase and regulation of glycogen biosynthesis in rat brain. J. biol. Chem. **244**, 3053–3061 (1969).

GOLDSTEIN, M., ANAGNOSTE, B., SHIRRON, C.: The effect of trivastal, haloperidol, and dibutyryl cyclic AMP on (^{14}C) dopamine synthesis in rat striatum. J. Pharm. Pharmacol. **25**, 348–351 (1973).

GOODMAN, D.B.P., RASMUSSEN, H., DIBELLA, F., GOTHROW, C.E.: Cyclic adenosine 3′:5′-monophosphate-stimulated phosphorylation of isolated neurotubule subunits. Proc. nat. Acad. Sci. (Wash.) **67**, 652–659 (1970).

GOODMAN, F.R., WEISS, G.B.: Dissociation by lanthanum of smooth muscle responses to potassium and acetylcholine. Amer. J. Physiol. **220**, 759–766 (1971).

GREENGARD, P., KEBABIAN, J.W.: Role of cyclic AMP in synaptic transmission in the mammalian peripheral nervous system. Fed. Proc. **33**, 1059–1068 (1974).

GREENGARD, P., KEBABIAN, J.W., MCAFEE, D.A.: Studies on the role of cyclic AMP in neural function. In: Cellular Mechanisms. Basel: S. Karger 1973.

GUIDOTTI, A., CHENEY, D.L., TRABUCCHI, M., DOTEUCHI, M., WANG, C.: Focussed microwave radiation: a technique to minimize post-mortem changes of cyclic nucleotides, DOPA and choline and to preserve brain morphology. Neuropharmacology **13**, 1115–1122 (1974).

GUIDOTTI, A., COSTA, E.: Involvement of adenosine 3′,5′-monophosphate in the activation of tyrosine hydroxylase elicited by drugs. Science **179**, 902–904 (1973).

GUIDOTTI, A., COSTA, E.: A role for nicotinic receptors in the regulation of the adenylate cyclase of adrenal medulla. J. Pharmacol. exp. Ther. **189**, 665–675 (1974).

GUIDOTTI, A., KUROSAWA, A., CHUANG, D.M., COSTA, E.: Protein kinase activation as an early event in the trans-synaptic induction of tyrosine-3-mono-oxygenase in adrenal medulla. Proc. nat. Acad. Sci. (Wash.) (in press).

GUIDOTTI, A., ZIVKOVIC, B., PFEIFFER, R., COSTA, E.: Involvement of 3′,5′-cyclic adenosine monophosphate in the increase of tyrosine hydroxylase activity elicited by cold exposure. Naunyn-Schmiedeberg's Arch. Pharmacol. **278**, 195–206 (1973).

GUNAGA, K.P., MENON, K.M.J.: Effect of catecholamines and ovarian hormones on cyclic AMP accumulation in rat hypothalamus. Biochem. biophys. Res. Commun. **54**, 440–448 (1973).

HAMPRECHT, B., SCHULTZ, J.: Stimulation by prostaglandin E_1 of adenosine 3′,5′-cyclic monophosphate formation in neuroblastoma cells in the presence of phosphodiesterase inhibitors. FEBS Letters **34**, 85–89 (1973).

HANNA, P.E., O'DEA, R.F., GOLDBERG, N.D.: Phosphodiesterase inhibition by papaverine and structurally related compounds. Biochem. Pharmacol. **21**, 2266–2268 (1972).

HARDMAN, J.G., SUTHERLAND, E.W.: Guanyl cyclase, an enzyme catalyzing the formation of guanosine 3′,5′-monophosphate from guanosine triphosphate. J. biol. Chem. **244**, 6363–6370 (1969).

HARRIS, J.E., MORGENROTH, V.H., ROTH, R.H., BALDESSARINI, R.J.: Regulation of catecholamine synthesis in the rat brain in vitro by cyclic AMP. Nature (Lond.) **252**, 156–158 (1974).

HAX, W.M.A., VAN VENROOIJ, G.E.P.M., VOSSENBERG, J.B.J.: Cell communication: a cyclic AMP mediated phenomenon. J. Membrane Biol. **19**, 253–266 (1974).

HAYAISHI, O., GREENGARD, P., COLOWICK, S.B.: On the equilibrium of the adenylate cyclase reaction. J. biol. Chem. **246**, 5840–5843 (1971).

HAYLETT, D.G., JENKINSON, D.H.: Effects of noradrenaline on the membrane potential and ionic permeability of parenchymal cells in the liver of the guinea-pig. Nature (Lond.) **224**, 80–81 (1969).

HEISLER, S., FAST, D., TENENHOUSE, A.: Role of Ca^{2+} and cyclic AMP in protein secretion from rat exocrine pancreas. Biochim. biophys. Acta (Amst.) **279**, 561–572 (1972).

HERMAN, Z.S.: Behavioral effects of dibutyryl cyclic 3′,5′ AMP, noradrenaline and cyclic 3′,5′ AMP in rats. Neuropharmacology **12**, 705–709 (1973).

HIDAKA, T., KURIYAMA, H.: Effects of catecholamines on the cholinergic neuromuscular transmission in fish red muscle. J. Physiol. (Lond.) **201**, 61–71 (1969).

HILL, R.G., SIMMONDS, M.A.: A method for comparing the potencies of γ-aminobutyric acid antagonists on single cortical neurones using microiontophoretic techniques. Brit. J. Pharmacol. **48**, 1–11 (1973).

HO, I.K., LOH, H.H., WAY, E.L.: Cyclic adenosine monophosphate antagonism of morphine analgesia. J. Pharmacol. exp. Ther. **185**, 336–346 (1973a).

HO, I.K., LOH, H.H., WAY, E.L.: Effects of cyclic 3′,5′-adenosine monophosphate on morphine tolerance and physical dependence. J. Pharmacol. exp. Ther. **185**, 347–357 (1973b).

HÖKFELT, T., UNGERSTEDT, U.: Specificity of 6-hydroxydopamine induced degeneration of central monoamine neurons: electron and fluorescence microscopic study with special reference to intracerebral injection of the nigro striatal dopamine system. Brain Res. **60**, 269–298 (1973).

HOFFER, B.J., SIGGINS, G.R., BLOOM, F.E.: Prostaglandins E_1 and E_2 antagonize norepinephrine effects on cerebellar Purkinje cells: Microelectrophoretic study. Science **166**, 1418–1420 (1969).

HOFFER, B.J., SIGGINS, G.R., BLOOM, F.E.: Studies on norepinephrine-containing afferents to Purkinje cells of rat cerebellum. II. Sensitivity of Purkinje cells to norepinephrine and related substances administered by microiontophoresis. Brain Res. **25**, 523–534 (1971a).

HOFFER, B.J., SIGGINS, G.R., OLIVER, A.P., BLOOM, F.E.: Cyclic AMP mediation of norepinephrine inhibition in rat cerebellar cortex: A unique class of synaptic responses. Ann. N.Y. Acad. Sci. **185**, 531–549 (1971b).

HOFFER, B.J., SIGGINS, G.R., OLIVER, A.P., BLOOM, F.E.: Cyclic adenosine monophosphate mediated adrenergic synapses to cerebellar Purkinje cells. In: Advances in Cyclic Nucleotide Research, vol. 1, 411–423 (1972).

HOFFER, B.J., SIGGINS, G.R., OLIVER, A.P., BLOOM, F.E.: Activation of the pathway from locus coeruleus to rat cerebellar Purkinje neurons: pharmacological evidence of noradrenergic central inhibition. J. Pharmacol. exp. Ther. **184**, 553–569 (1973).

HOFFER, B.J., SIGGINS, G.R., WOODWARD, D.J., BLOOM, F.E.: Spontaneous discharge of Purkinje neurons after destruction of catecholamine-containing afferents by 6-hydroxydopamine. Brain Res. **30**, 425–430 (1971).

HOFMANN, F., SOLD, G.: A protein kinase activity from rat cerebellum stimulated by guanosine 3′,5′-monophosphate. Biochem. biophys. Res. Commun. **49**, 1100–1107 (1972).

HONDA, F., IMAMURA, H.: Inhibition of cyclic 3′,5′-nucleotide phosphodiesterase by phenothiazine and reserpine derivatives. Biochim. biophys. Acta (Amst.) **161**, 267–269 (1968).

HOROWITZ, J.M., HORWITZ, B.A., SMITH, R.E.: Effect in vivo of norepinephrine on the membrane resistance of brown fat cells. Experientia (Basel) **27**, 1419 (1971).

HORWITZ, B.A., HOROWITZ, J.M., SMITH, R.E.: Norepinephrine-induced depolarization of brown fat cells. Proc. nat. Acad. Sci. (Wash.) **64**, 113–120 (1969).

HOWELL, S.L., WHITFIELD, M.: Localization of adenyl cyclase in islet cells. J. Histochem. Cytochem. **20**, 873 (1972).

HUANG, M., GRUENSTEIN, E., DALY, J.W.: Depolarizing-evoked accumulation of cyclic AMP in brain slices: inhibition by exogenous adenosine deaminase. Biochim. biophys. Acta (Amst.) **329**, 147–151 (1973a).

HUANG, M., HO, A.K.S., DALY, J.W.: Accumulation of adenosine cyclic 3′,5′-monophosphate in rat cerebral cortical slices. Stimulatory effects of alpha and beta adrenergic agents after treatment with 6-hydroxydopamine, 2,3,5-trihydroxyphenethylamine and dehydroxytryptamines. Molec. Pharmacol. **9**, 711–717 (1973b).

HUANG, M., SHIMIZU, H., DALY, J.: Regulation of adenosine cyclic 3′,5′-phosphate formation in cerebral cortical slices. Interaction among norepinephrine, histamine, serotonin. Molec. Pharmacol. **7**, 155–162 (1971).

HUXLEY, A.F.: Muscular contraction. J. Physiol. (Lond.) **243**, 1–44 (1974).

HUXLEY, J.S.: Chemical regulation and the hormone concept. Biol. Rev. **10**, 427–441 (1935).

ITO, Y., KURIYAMA, H., TASHIRO, N.: Effects of catecholamines on the neuromuscular junction of the somatic muscle of the earthworm pheretima communissima. J. exp. Biol. **54**, 167–186 (1971).

JACOBOWITZ, D.: Catecholamine fluorescence studies of adrenergic neurons and chromaffin cells in sympathetic ganglia. Fed. Proc. **29**, 1929–1944 (1970).

JANIEC, W., TRZECIAK, H., HERMAN, Z.: The influence of adrenaline and optical isomers of INPEA on the adenyl cyclase activity in main hemispheres of rats in vitro. Arch. int. Pharmacodyn **185**, 254–258 (1970).

JENKINSON, D.H., STAMENOVIC, B.A., WHITAKER, B.D.L.: The effect of noreadrenaline on the end-plate potential in twitch fibres of the frog. J. Physiol. (Lond.) **195**, 743–754 (1968).

JOHNSON, E.M., MAENO, H., GREENGARD, P.: Phosphorylation of endogenous protein of rat brain by a cyclic adenosine 3′,5′-monophosphate dependent protein kinase. J. biol. Chem. **246**, 7731–7739 (1971).

JOHNSON, E.M., UEDA, T., MAENO, H., GREENGARD, P.: Adenosine 3′,5′-monophosphate-dependent phosphorylation of a specific protein in synaptic membrane fractions from rat cerebrum. J. biol. Chem. **247**, 5650–5652 (1972).

JORDAN, L.M., LAKE, N., PHILLIS, J.W.: Mechanism of noradrenaline depression of cortical neurones: a species comparison. Europ. J. Pharmacol. **20**, 381–384 (1972).

JOUVET, M.: The role of monoamines and acetylcholine containing neurons in the regulation of the sleep-waking cycle. Ergebn. Physiol. **64**, 168–307 (1972).

KADLEC, O., MASEK, K., SEFERNA, I.: The effect of papaverine on $^{45}Ca^{2+}$ uptake in partly depolarized taeni coli of the guinea-pig. J. Pharm. Pharmacol. **25**, 914–916 (1973).

KAKIUCHI, K.S., MARKS, B.H.: Adrenergic effects on pineal cell membrane potential. Life Sci. **11**, 285–291 (1972).

KAKIUCHI, S.: Cyclic 3′,5′-nucleotide phosphodiesterase of rat brain and other tissues: regulation of activity by Ca^{2+} and the modulator protein. In: Cellular Mechanisms. Basel: S. Karger 1973.

KAKIUCHI, S.: Ca^{2+} plus Mg^{2+}-dependent phosphodiesterase and its modulator from various rat tissues. In: Advances in Cyclic Nucleotide Research, vol. 5. New York: Raven Press (in press).

KAKIUCHI, S., RALL, T.W.: Studies on adenosine 3′,5′-phosphate in rabbit cerebral cortex. Molec. Pharmacol. **4**, 379–388 (1968a).

KAKIUCHI, S., RALL, T.W.: The influence of chemical agents on the accumulation of adenosine 3′,5′-phosphate in slices of rabbit cerebellum. Molec. Pharmacol. **4**, 367–378 (1968b).

KAKIUCHI, S., RALL, T.W., MCILWAIN, H.: The effect of electrical stimulation upon the accumulation of adenosine 3′,5′-phosphate in isolated cerebral tissue. J. Neurochem. **16**, 485–491 (1969).

KAKIUCHI, S., YAMAZAKI, R.: Calcium dependent phosphodiesterase activity and its activating factor (PAF) from brain. Studies on cyclic 2′,5′-nucleotide phosphodiesterase. Biochem. biophys. Res. Commun. **41**, 1104–1110 (1970).

KAKIUCHI, S., YAMAZAKI, R., TESHIMA, Y.: Cyclic 3′,5′-nucleotide phosphodiesterase. IV. Two enzymes with different properties from brain. Biochem. biophys. Res. Commun. **42**, 968–974 (1971).

KAKIUCHI, S., YAMAZAKI, R., TESHIMA, Y., UENISHI, K.: Regulation of nucleoside cyclic 3′,5′-monophosphate phosphodiesterase activity from rat brain by a modulator and Ca^{2+}. Proc. nat. Acad. Sci. (Wash.) **70**, 3526–3535 (1973).

KAKIUCHI, S., YAMAZAKI, R., TESHIMA, Y., UENISHI, K., MIYAMOTO, E.: Multiple cyclic nucleotide phosphodiesterase activities from rat tissues and occurrence of a calcium-plus-magnesium-ion-dependent phosphodiesterase and its protein activator. Biochem. J. **146**, 109–120 (1975).

KALIX, P., MCAFEE, D.A., CHORDERET, M., GREENGARD, P.: Pharmacological analysis of synaptically mediated increase in cyclic AMP in rabbit superior cervical ganglion. J. Pharmacol. exp. Ther. **188**, 676–687 (1974).

KANDEL, E.R.: Dale's principle and the functional specificity of neurons. In: Psychopharmacology – A Ten Year Progress Report. Washington, D.C.: U.S. Govt. Printing Office 1968.

KAROBATH, M., LEITICH, H.: Antipsychotic drugs and dopamine-stimulated adenylate cyclase prepared from corpus striatum of rat brain. Proc. nat. Acad. Sci. (Wash.) **71**, 2915–2918 (1974).

KATZ, A.M., TADA, M., REPKE, D.J., IORIO, J.M., KRICHBERGER, M.A.: Adenylate cyclase: Its probable localization in sarcoplasmic reticulum as well as sarcolemma of the canine heart. J. Mol. Cell. Cardiol. **6**, 73–78 (1974).

KATZ, B.: Nerve, Muscle, and Synapse. New York: McGraw-Hill 1966.

KATZ, S., TENENHOUSE, A.: The relationship of adenyl cyclase to the activity of other ATP utilizing enzymes and phosphodiesterase in preparations of rat brain. Mechanism of stimulation of cyclic AMP accumulation by adrenaline, ouabain, and Mg^{++}. Brit. J. Pharmacol. **48**, 516–526 (1973).

KATZMAN, R., BJÖRKLUND, A., OWMAN, C., STENEVI, U., WEST, K.: Evidence for regenerative axon sprouting of central catecholamine neurons in rat mesencephalon following electrolytic lesions. Brain Res. **25**, 579–596 (1971).

KAUFFMAN, F.E., HARKONEN, M.H.A., JOHNSON, E.E.: Adenyl cyclase and phosphodiesterase activity in cerebral cortex of normal and undernourished neonatal rats. Life Sci. **11**, 613–621 (1972).

KAUKEL, E., HILZ, H.: Permeation of dibutyryl cAMP into hela cells and its conversion to monobutyryl cAMP. Biochem. biophys. Res. Commun. **46**, 1011–1018 (1972).

KEBABIAN, J.W., GREENGARD, P.: Dopamine-sensitive adenyl cyclase: possible role in synaptic transmission. Science **174**, 1346–1349 (1971).

KEBABIAN, J.W., PETZOLD, G.L., GREENGARD, P.: Dopamine-sensitive adenylate cyclase in caudate nucleus of rat brain and its similarities to the dopamine receptor. Proc. nat. Acad. Sci. (Wash.) **69**, 2145–2150 (1972).

KEEN, P., MCLEAN, W.G.: Effect of dibutyryl-cyclic AMP and dexamethasone on noradrenaline synthesis in isolated superior cervical ganglia. J. Neurochem. **22**, 5–10 (1974).

KIMURA, H., THOMAS, E., MURAD, F.: Effects of decapitation, ether and pentobarbital on guanosine 3′-5′ phosphate and adenosine 3′-5′ phosphate levels in rat tissues. Biochim. biophys. Acta (Amst.) **343**, 519–528 (1974).

KIRCHBERGER, M.A., TADA, M., KATZ, A.M.: Adenosine 3′,5′-monophosphate-dependent protein kinase-catalyzed phosphorylation reaction and its relationship to calcium transport in cardiac sarcoplasmic reticulum. J. biol. Chem. **249**, 6166–6173 (1974).

KLAINER, L.M., CHI, Y-M., FRIEDBERG, S.L., RALL, T.W., SUTHERLAND, E.: Adenyl cyclase. IV. The effects of neurohormones on the formation of adenosine 3′-5′ phosphate by preparations from brain and other tissues. J. biol. Chem. **237**, 1239–1243 (1962).

KLAWANS, H.L., MOSES, H., BEAULIEU, D.M.: The influence of caffeine on d-amphetamine- and apomorphine-induced stereotyped behavior. Life Sci. **14**, 1493–1500 (1974).

KLEIN, D.C., YUWILER, A., WELLER, J.L., PLOTKIN, S.: Postsynaptic adrenergic-cyclic AMP control of the serotonin content of cultured rat pineal glands. J. Neurochem. **21**, 1261–1271 (1973).

KNULL, H.R., TAYLOR, W.F., WELLS, W.W.: Insulin effects on brain energy metabolism and the related hexokinase distribution. J. biol. Chem. **249**, 6930–6935 (1974).

KODAMA, T., MATSUKO, Y., SHIMIZU, H.: The cyclic AMP system of human brain. Brain Res. **50**, 135–146 (1973).

KOELLE, G.B., FRIEDENWALD, J.S.: A histochemical method for localizing cholinesterase activity. Proc. Soc. exp. Biol. (N.Y.) **70**, 617–622 (1949).

KRISHNA, G., DITZION, B.R., GESSA, G.L.: Intense ergotrophic stimulation induced by intracerebral injection of dibutyryl cyclic 3′-5′ AMP. Proc. Int. Union Physiol. Sci., Washington, D.C. **1**, 247 (1968).

KRISHNA, G., MOSKOWITZ, J., DEMPSEY, P., BRODIE, B.B.: The effect of norepinephrine and insulin on brown fat cell membrane potentials. Life Sci. **9**, 1353–1361 (1970).

KRNJEVIĆ, K., DUMAIN, R., RENAUD, L.: The mechanism of excitation by acetylcholine in the cerebral cortex. J. Physiol. (Lond.) **215**, 247–268 (1971).

KROEGER, E.A., MARSHALL, J.M.: Beta-adrenergic effects on rat myometrium: mechanisms of membrane hyperpolarization. Amer. J. Physiol. **225**, 1339–1345 (1973).

KUBA, K.: Effects of catecholamines on the neuromuscular junction in the rat diaphragm. J. Physiol. (Lond.) **211**, 551–570 (1970).

KUBA, K., KOKETSU, K.: Ionic mechanism of the slow excitatory postsynaptic potential in bullfrog sympathetic ganglion cells. Brain Res. **81**, 338–342 (1974).

KUBA, K., TOMITA, T.: Noradrenaline action on nerve terminal in the rat diaphragm. J. Physiol. (Lond.) **217**, 19–31 (1971).

KUFFLER, S.W.: Transmitter mechanism at the nerve-muscle junction. Arch. Sci. physiol. **3**, 585–601 (1949).

KUO, J.F.: Guanosine 3':5'-monophosphate-dependent protein kinases in mammalian tissues. Proc. nat. Acad. Sci. (Wash.) **71**, 4037–4041 (1974).

KUO, J.F., GREENGARD, P.: Cyclic nucleotide-dependent protein kinases. IV. Widespread occurrence of adenosine 3',5'-monophosphate dependent protein kinase in various tissues and phyla of the animal kingdom. Proc. nat. Acad. Sci. (Wash.) **64**, 1349–1355 (1969a).

KUO, J.F., GREENGARD, P.: Adenosine 3',5'-monophosphate dependent protein kinase from brain. Science **165**, 63–65 (1969b).

KUO, J.F., GREENGARD, P.: Stimulation of adenosine 3',5'-monophosphate-dependent protein kinases by some analogs of adenosine 3',5'-monophosphate. Biochem. biophys. Res. Commun. **40**, 1032–1038 (1970a).

KUO, J.F., GREENGARD, P.: Cyclic nucleotide-dependent protein kinases. VII. Comparison of various histones as substrates for adenosine 3',5'-monophosphate-dependent and guanosine 3',5'-monophosphate-dependent protein kinases. Biochim. biophys. Acta (Amst.) **212**, 434–440 (1970b).

KUO, J.F., GREENGARD, P.: Stimulation of cyclic GMP dependent protein kinase by a protein fraction which inhibits cyclic AMP-dependent protein kinases. Fed. Proc. **30**, 1089 (1973).

KUO, J.F., LEE, T.P., REYES, P.L., WALTON, K.G., DONNELLY, T.E., GREENGARD, P.: Cyclic nucleotide-dependent protein kinases. X. An assay method for the measurement of guanosine 3',5'-monophosphate in various biological materials and a study of agents regulating its levels on heart and brain. J. biol. Chem. **247**, 16–22 (1972).

KUO, J.F., MIYAMOTO, E., REYES, P.L.: Activation and dissociation of adenosine 3',5'-monophosphate-dependent protein kinase by various cyclic nucleotide analogs. Biochem. Pharmacol. **23**, 2011–2021 (1974).

KURIYAMA, K., ISREAL, M.A.: Effect of ethanol administration on cyclic 3',5'-adenosine-monophosphate metabolism in brain. Biochem. Pharmacol. **22**, 2919–2922 (1973).

LAKE, N., JORDAN, L.M.: Failure to confirm cyclic AMP as second messenger for norepinephrine in rat cerebellum. Science **183**, 663–664 (1974).

LAKE, N., JORDAN, L.M., PHILLIS, J.W.: Mechanism of noradrenaline actions in cat cerebral cortex. Nature (Lond.) **240**, 249–250 (1972).

LAKE, N., JORDAN, L.M., PHILLIS, J.W.: Evidence against cyclic adenosine 3',5'-monophosphate (AMP) mediation of noradrenaline depression of cerebral cortical neurones. Brain Res. **60**, 411–421 (1973).

LANDIS, S.C., BLOOM, F.E.: Fluorescence and electron microscopic analysis of catecholamine containing fibers in mutant mouse cerebellum. 4th Annu. Meeting Soc. Neurosci. 297 1974.

LANGAN, T.: Protein kinases and protein kinase substrates. In: Advances in Cyclic Nucleotide Research, vol. 1. New York: Raven Press 1972.

LEE, T-P., KUO, J.F., GREENGARD, P.: Role of muscarinic cholinergic receptors in regulation of guanosine 3',5'-cyclic monophosphate content in mammalian brain, heart muscle, and intestinal smooth muscle. Proc. nat. Acad. Sci. (Wash.) **69**, 3287–3289 (1972).

LEFKOWITZ, R.J.: Stimulation of catecholamine-sensitive adenylate by 5'-guanylyl imido-diphosphate. J. biol. Chem. **249**, 6119–6124 (1974).

LEFKOWITZ, R.J., MUKHERJEE, C., COVERSTONE, M., CARON, M.G.: Stereospecific (^{3}H) (–) – alprenolol binding sites, β-adrenergic receptors and adenylate cyclase. Biochem. biophys. Res. Commun. **60**, 703–709 (1974).

LENTZ, T.L.: A role of cyclic AMP in a neurotrophic process. Nature (Lond.) **238**, 154–155 (1972).

LEVITAN, I.B., BARONDES, S.H.: Octopamine- and serotonin-stimulated phosphorylation of specific protein in the abdominal ganglion of Aplysia californica. Proc. nat. Acad. Sci. (Wash.) **72**, 1145–1148 (1974).

LEVITAN, I.B., MADSEN, C.J., BARONDES, S.H.: Cyclic AMP and amine effects on phosphorylation of specific protein in abdominal ganglion of aplysia californica: localization and kinetic analysis. J. Neurobiol. **5**, 475–588 (1974).

LIBET, B.: Generation of slow inhibitory and excitatory postsynaptic potentials. Fed. Proc. **29**, 1945–1949 (1970).

LIBET, B., KOBAYASHI, H.: Generation of adrenergic and cholinergic potentials in sympathetic ganglion cells. Science **164**, 1530–1532 (1969).

LIN, Y.M., LIU, Y.P., CHEUNG, W.Y.: Cyclic 3′:5′-nucleotide phosphodiesterase. Purification, characterization, and active form of the protein activator from bovine brain. J. biol. Chem. **249**, 4943–4954 (1974).

LINDL, T., CRAMER, H.: Formation, accumulation and release of adenosine 3′,5′-monophosphate induced by histamine in the superior cervical ganglion of the rat in vivo. Biochim. biophys. Acta (Amst.) **343**, 182–191 (1974).

LINDVALL, O., BJÖRKLUND, A.: The organization of the ascending catecholamine neuron systems in the rat brain. Acta physiol. scand. **412**, 1–48 (1974).

LINDVALL, O., BJÖRKLUND, A., MOORE, R.Y., STENEVI, U.: Mesencephalic dopamine neurons projecting to neocortex. Brain Res. **81**, 325–331 (1974).

LING, G., GERARD, R.W.: Normal membrane potential of frog sartorius fibers. J. cell. comp. Physiol. **34**, 383–394 (1949).

LIPKIN, D., COOK, W.H., MARKHAM, R.: Adenosine-3′:5′-phosphoric acid: A proof of structure. J. Amer. chem. Soc. **81**, 6198–6203 (1959).

LOEWI, O.: Über humorale Übertragbarkeit der Herznervenwirkung. Pflügers Arch. ges. Physiol. **189**, 239–242 (1921).

LONDOS, C., SOLOMON, Y., LIN, M.C., HARWOOD, J.P., SCHRAMM, M., WOLFF, J., RODBELL, M.: 5-Guanylyl-imidodiphosphate, a potent activator of adenylate cyclase systems in eukaryotic cells. Proc. nat. Acad. Sci. (Wash.) **71**, 3087–3090 (1974).

LORENZO, R. DE, GREENGARD, P.: Activation by adenosine 3′:5′-monophosphate of a membrane bound phosphoprotein phosphatase from toad bladder. Proc. nat. Acad. Sci. (Wash.) **70**, 1831–1835 (1973).

LORENZO, R. DE, WALTON, K.G., CURRAN, P.F., GREENGARD, P.: Regulation of phosphorylation of a specific protein in toad bladder membrane by antidiuretic hormone and cyclic AMP, and its possible relationship to membrane permeability changes. Proc. nat. Acad. Sci. (Wash.) **70**, 880–884 (1973).

LOWRY, O.H., PASSONNEAU, J.V., HASSELBERGER, F.X., SCHULTZ, D.W.: Effect of ischemia on known substrates and cofactors of the glycolytic pathway in brain. J. biol. Chem. **239**, 18–30 (1964).

LUST, W.D., PASSONNEAU, J.V., VEECH, R.L.: Cyclic adenosine monophosphate metabolites and phosphorylase in neuronal tissue. A comparison of methods of fixation. Science **181**, 280–282 (1973).

MACKAY, A.V.P., IVERSEN, L.L.: Increased tyrosine hydroxylase activity of sympathetic ganglia cultured in the presence of dibutyryl cyclic AMP. Brain Res. **48**, 424–426 (1972).

MAEDA, T., TOHYAMA, M., SHIMIZU, N.: Modification of postnatal development of neocortex in rat brain with experimental deprivation of locus coeruleus. Brain Res. **70**, 515–520 (1974).

MAENO, H., GREENGARD, P.: Phosphoprotein phosphatases from rat cerebral cortex. J. biol. Chem. **247**, 3269–3277 (1972).

MAENO, H., JOHNSON, E.M., GREENGARD, P.: Subcellular distribution of adenosine 3′,5′-monophosphate-dependent protein kinase in rat brain. J. biol. Chem. **246**, 134–142 (1971).

MAENO, H., UEDA, T., GREENGARD, P.: Adenosine 3′:5′-monophosphate dependent protein phosphatase activity in synaptic membrane fractions. J. Cyclic Nucleotide Res. **1**, 37–48 (1975).

MAGARIBUCHI, M., KURIYAMA, H.: Effects of noradrenaline and isoprenaline on the electrical and mechanical activities of guinea pig depolarized taenia coli. Jap. J. Physiol. **22**, 253–270 (1972).

MAGUN, B.: Two actions of cyclic AMP on melanosome movement in frog skin. J. Cell Biol. **57**, 845–858 (1973).

MAO, C.C., GUIDOTTI, A., COSTA, E.: Inhibition by diazepam of the tremor and the increase of cerebellar cGMP content elicited by harmaline. Brain Res. **83**, 526–529 (1974a).

MAO, C.C., GUIDOTTI, A., COSTA, E.: Interactions between γ-amino-bytyric acid and guanosine cyclic 3′-5′ monophosphate in rat cerebellum. Molec. Pharmacol. **10**, 736–745 (1974b).

MAO, C.C., GUIDOTTI, A., COSTA, E.: The regulation of cyclic guanosine monophosphate in rat cerebellum: possible involvement of putative amino acid neurotransmitters. Brain Res. **79**, 510–514 (1974c).

MATTHEWS, E.K., SAFFRAN, M.: Ionic dependence of adrenal steroidogenesis and ACTH-induced changes in the membrane potential of adreno-cortical cells. J. Physiol. (Lond.) **234**, 43–64 (1973).

MAYER, S.E.: Effect of catecholamines on cardiac metabolism. Circulat. Res. Suppl. III **34–35**, 129–137 (1974).

McAFEE, D.A., GREENGARD, P.: Adenosine 3′,5′-monophosphate: electrophysiological evidence for a role in synaptic transmission. Science **178**, 310–312 (1972).

McAFEE, D.A., SCHORDERET, M., GREENGARD, P.: Adenosine 3′,5′-monophosphate in nervous tissue: increase associated with synaptic transmission. Science **171**, 1156–1158 (1971).

McCUNE, R.W., GILL, T.H., VON HUNGEN, K., ROBERTS, S.: Catecholamine sensitive adenyl cyclase in cell-free preparations from rat cerebral cortex. Life Sci. **10**, 443–450 (1971).

McMANUS, J.P., WHITFIELD, J.F.: Cyclic AMP, prostaglandins and the control of cell proliferation. Prostaglandins **6**, 475–487 (1974).

MEINERTZ, T., NAWRATH, H., SCHOLZ, H., WINTER, K.: Effect of DB-c-AMP on mechanical characteristics of ventricular and atrial preparations of several mammalian species. Naunyn-Schmiedeberg's Arch. Pharmacol. **282**, 143–153 (1974).

MEYER, R.B. JR., MILLER, J.P.: Analogs of cyclic AMP and cyclic GMP: general methods of synthesis and the relationship of structure to enzymic activity. Life Sci. **14**, 1019–1040 (1974).

MIKI, N., KEIRNS, J.J., MARKUS, F.R., FREEMAN, J., BITENSKY, M.W.: Regulation of cyclic nucleotide concentrations in photoreceptors: An ATP-dependent stimulation of cyclic nucleotide phosphodiesterase by light. Proc. nat. Acad. Sci. (Wash.) **70**, 3820–3824 (1973).

MILLER, C.A., LEVINE, E.M.: Neuroblastoma: synchronization of neurite growth in cultures grown in collagen. Science **177**, 799–801 (1972).

MILLER, J.P., BOSWELL, K.H., MUNEYANA, K., SIMON, L.N., ROBINS, R.K., SHUMAN, D.A.: Synthesis and biochemical studies of various 8-substituted derivatives of guanosine 3′,5′-cyclic phosphate, inosine 3′,5′-cyclic phosphate and xanthosine 3′,5′-cyclic phosphate. Biochemistry (Wash.) **12**, 5310–5319 (1973).

MILLER, R.J., HORN, A.S., IVERSEN, L.L.: The action of neuroleptic drugs on dopamine-stimulated adenosine cyclic 3,5′-monophosphate production in rat neostriatum and limbic forebrain. Molec. Pharmacol. **10**, 759–766 (1974).

MILLER, W.H.: Cyclic nucleotides and photoreception. Exp. Eye Res. **16**, 357–363 (1973).

MINOR, A.V., SAKINA, N.L.: Role of cyclic adenosine 3′-5′ monophosphate in olfactory reception. Neurofizilogia **2**, 415–422 (1973).

MISHRA, R.K., GARDNER, E.L., KATZMAN, R., MAKMAN, M.H.: Enhancement of dopamine-stimulated adenylate cyclase activity in rat caudate after lesions in substantia nigra: evidence for denervation supersensitivity. Proc. nat. Acad. Sci. (Wash.) **71**, 3883–3887 (1974).

MITZNEGG, P., SCHUBERT, E., HEIM, F.: The influence of low and high doses of theophylline on spontaneous motility and cyclic 3′,5′ AMP content in isolated rat uterus. Life Sci. **14**, 711–717 (1974).

MIYAMOTO, E., KAKIUCHI, S.: In Vitro and In Vivo phosphorylation of myelin basic protein by exogenous and endogenous adenosine 3′5′-monophosphate-dependent protein kinases in brain. J. biol. Chem. **249**, 2569–2777 (1974).

MIYAMOTO, E., KUO, J.F., GREENGARD, P.: Adenosine 3′,5′-monophosphate-dependent protein kinase from brain. Science **165**, 63–65 (1969a).

MIYAMOTO, E., KUO, J.F., GREENGARD, P.: Cyclic nucleotide-dependent protein kinases. I. Purification and properties of adenosine 3′,5′-monophosphate-dependent protein kinase from bovine brain. J. biol. Chem. **244**, 6395–6402 (1969b).

MIYAMOTO, E., PETZOLD, G.L., HARRIS, J.S., GREENGARD, P.: Dissociation and concomitant activation of adenosine 3′,5′-monophosphate-dependent protein kinase by histone. Biochem. biophys. Res. Commun. **44**, 305–312 (1971).

MOLINOFF, P.B., AXELROD, J.: Biochemistry of catecholamines. Ann. Rev. Biochem. **40**, 465–500 (1971).

MONN, E., CHRISTIANSEN, R.O.: Adenosine 3′,5′-monophosphate phosphodiesterase: multiple molecular forms. Science **173**, 540–541 (1971).

MOORE, R.Y., BJÖRKLUND, A., STENEVI, U.: Plastic changes in the adrenergic innervation of the rat septal area in response to denervation. Brain Res. **33**, 13–35 (1971).

MORGENROTH, V.H. III., BOADLE-BIBER, M., ROTH, R.H.: Tyrosine hydrolase: Activation by nerve stimulation. Proc. nat. Acad. Sci. (Wash.) **71**, 4283–4287 (1974).

MOSES, H.L., ROSENTHAL, A.S.: Pitfalls in the use of lead ion for histochemical localization of nucleoside phosphatases. J. Histochem. Cytochem. **16**, 530–539 (1968).

MUELLER, R.A., OTTEN, U., THOENEN, H.: The role of adenosine cyclic 3′,5′-monophosphate in reserpine-initiated adrenal medullary tyrosine hydroxylase induction. Molec. Pharmacol. **10**, 855–860 (1974).

MUELLER, R.A., THOENEN, H., AXELROD, J.: Increase in tyrosine hydroxylase activity after reserpine administration. J. Pharmacol. exp. Ther. **169**, 74–79 (1969).

MULLEROE, B., SCHWABE, B.: Role of cyclic AMP for function of peripheral and central nervous system. Fortsch. Neurol. Psychiat. **41**, 509–526 (1973).

MURAD, F., MANGANIELLO, V., VAUGHAN, M.: A simple sensitive protein binding assay for guanosine 3′-5′ monophosphate. Proc. nat. Acad. Sci. (Wash.) **68**, 736–739 (1971).

MURPHY, D.L., DONNELLY, C., MOSKOWITZ, J.: Inhibition by lithium of prostaglandin E_1 and norepinephrine effects on cyclic adenosine monophosphate production in human platelets. Clin. Pharmacol. Ther. **14**, 810–814 (1973).

NAHORSKI, S.R., ROGERS, K.J., PINNS, J.: Cerebral phosphodiesterase and dopamine receptor. J. Pharm. Pharmacol. **25**, 912–913 (1973).

NAITO, K., KURIYAMA, K.: Effect of morphine administration on adenyl cyclase and 3′,5′-cyclic nucleotide phosphodiesterase activities in the brain. Jap. J. Pharmacol. **23**, 274–276 (1973).

NAKAZAWA, K., SANO, M.: Studies on guanylate cyclase. A new assay method for guanylate cyclase and properties of the cyclase from rat brain. J. biol. Chem. **249**, 4207–4211 (1974).

NATHANSON, J.A., GREENGARD, P.: Octopamine-sensitive adenylate cyclase: evidence for a biological role of octopamine in nervous tissue. Science **180**, 308–310 (1973).

NATHANSON, J.A., GREENGARD, P.: Serotonin-sensitive adenylate cyclase in neural tissue and its similarity to the serotonin receptor: a possible site of lysergic acid diethylamide. Proc. nat. Acad. Sci. (Wash.) **71**, 797–801 (1974).

NEELON, F.A., BIRCH, B.M.: Cyclic adenosine 3′:5′-monophosphate-dependent protein kinase. J. biol. Chem. **248**, 8361–8365 (1973).

NELSON, C.N., HOFFER, B.J., CHU, N-S., BLOOM, F.E.: Cytochemical and pharmacological studies on polysensory neurons in the primate frontal cortex. Brain Res. **62**, 115–133 (1973).

NORBERG, K.A., RITZEN, M., UNGERSTEDT, U.: Histochemical studies on a special catecholamine-containing cell type in sympathetic ganglia. Acta physiol. scand. **67**, 260–270 (1966).

OBATA, K., TAKEDA, K., SHINOZAKI, H.: Electrophoretic release of γ-aminobutyric acid and glutamic acid from micropipettes. Neuropharmacol. **9**, 191–194 (1970).

OBATA, K., YOSHIDA, M.: Caudate-evoked inhibition and actions of GABA and other substances on cat pallidal neurons. Brain Res. **64**, 455–459 (1973).

OLIVER, A.P., SEGAL, M.: Transmembrane changes in hippocampal neurons: hyperpolarizing actions of norepinephrine, cyclic AMP, and locus coeruleus. Proc. Soc. Neurosci. 361 (1974).

OLSON, L., FUXE, K.: On the projections from the locus coeruleus noradrenaline neurons. Brain Res. **28**, 165–171 (1971).

OLSON, L., FUXE, K.: Further mapping out of central noradrenaline nervous systems: Projections of the subcoeruleus area. Brain Res. **43**, 289–295 (1972).

OTTEN, U., MUELLER, R.A., OESCH, F., THOENEN, H.: Location of an isoproterenol-responsive cyclic AMP pool in adrenergic nerve cell bodies and its relationship to tyrosine 3-monooxygenase induction. Proc. nat. Acad. Sci. (Wash.) **71**, 2217–2221 (1974).

OTTEN, U., OESCH, F., THOENEN, H.: Dissociation between changes in cyclic AMP and subsequent induction of tyrosine hydroxylase in rat superior cervical ganglion and adrenal medulla. Naunyn-Schmiedeberg's Arch. Pharmacol. **280**, 129–140 (1973).

PALAY, S.L., CHAN-PALAY, V.: Cerebellar Cortex. Cytology and Organization. Berlin-Heidelberg-New York: Springer 1974.

PALMER, G.C.: Increased cyclic AMP response to norepinephrine in the rat brain following 6-hydroxydopamine. Neuropharmacol. **11**, 145–149 (1972).

PALMER, G.C.: Adenyl cyclase in neuronal and glial-enriched fractions from rat and rabbit brain. Res. Commun. Chem. Path. Pharmacol. **5**, 603–613 (1973a).

PALMER, G.C.: Influence of amphetamines, protriptyline and pargyline on the time course of the norepinephrine-induced accumulation of cyclic AMP in rat brain. Life Sci. **12**, 345–355 (1973b).

PALMER, G.C., BURKS, T.F.: Central and peripheral adrenergic blocking actions of LSD and BOL. Europ. J. Pharmacol. **16**, 113–116 (1971).

PALMER, G.C., ROBISON, G.A., MANIAN, A.A., SULSER, F.: Modification by psychotropic drugs of the cyclic AMP response to norepinephrine in the rat brain in vitro. Psychopharmacologia (Berl.) **23**, 201–211 (1972a).

PALMER, G.C., SCHMIDT, M.J., ROBISON, G.A.: Development and characteristics of the histamine-induced accumulation of cyclic AMP in the rabbit cerebral cortex. J. Neurochem. **19**, 2251–2256 (1972b).

PALMER, G.C., SULSER, F., ROBISON, G.A.: Effects of neurohumoral and adrenergic agents on cyclic AMP levels in various areas of the rat brain in vitro. Neuropharmacol. **12**, 327–337 (1973).

PANNBACKER, R.G.: Control of guanylate cyclase activity in the rod outer segment. Science **182**, 1138–1139 (1973).

PANNBACKER, R.G., FLEISCHMAN, D.E., REED, D.W.: Cyclic nucleotide phosphodiesterase: high activity in a mammalian photoreceptor. Science **175**, 757–758 (1972).

PAPAHADJOPOULOS, D., POSTE, G., MAYHEW, E.: Cellular uptake of cyclic AMP captured within phospholipid vesicles and effect on cell growth behavior. Biochim. biophys. Acta (Amst.) **363**, 404–418 (1974).

PATON, W.D.M.: Central and synaptic transmission in the nervous system (pharmacological aspects). Ann. Rev. Pharmacol. **20**, 431–462 (1958).

PAUL, M.I., CRAMER, H., BUNNEY, W.E. JR.: Urinary adenosine 3′,5′-monophosphate in the switch process from depression to mania. Science **171**, 300–303 (1971a).

PAUL, M.I., CRAMER, H., GOODWIN, F.K.: Urinary cyclic AMP excretion in depression and mania. Arch. gen. Psychiat. **24**, 327–333 (1971b).

PERKINS, J.P.: Adenyl cyclase. In: Advances in Cyclic Nucleotide Research, vol. 3. New York: Raven Press 1973.

PERKINS, J.P., MOORE, M.M.: Adenyl cyclase of rat cerebral cortex. J. biol. Chem. **246**, 62–68 (1971).

PERKINS, J.P., MOORE, M.M.: Regulation of the adenosine cyclic 3′5′-monophosphate content of rat cerebral cortex: ontogenetic development of the responsiveness to catecholamines and adenosine. Molec. Pharmacol. **9**, 774–782 (1973a).

PERKINS, J.P., MOORE, M.M.: Characterization of the adrenergic receptors mediating a rise in cyclic 3′,5′-adenosine monophosphate in rat cerebral cortex. J. Pharmacol. exp. Ther. **185**, 371–378 (1973b).

PETERSEN, O.H.: The effect of glucagon on the liver cell membrane potential. J. Physiol. (Lond.) **239**, 647–656 (1974).

PHILLIS, J.W.: Evidence for cholinergic transmission in the cerebral cortex. Advan. Behav. Biol. **10**, 57–80 (1974a).

PHILLIS, J.W.: The role of calcium in the central effects of biogenic amines. Life Sci. **14**, 1189–1201 (1974b).

PHILLIS, J.W., KOSTOPOULOS, G.K., LIMACHER, J.J.: A potent depressant action of adenine derivatives on cerebral cortical neurons. Europ. J. Pharmacol. **30**, 125–129 (1975).

PHILLIS, J.W., LAKE, N., YARBOROUGH, G.: Calcium mediation of the inhibitory effects of biogenic amines on cerebral cortical neurones. Brain Res. **53**, 465–469 (1973).

PICKEL, V.M., SEGAL, M., BLOOM, F.E.: A radioautographic study of the efferent pathways of the nucleus locus coeruleus. J. comp. Neurol. **155**, 15–42 (1974a).

PICKEL, V.M., SEGAL, M., BLOOM, F.E.: Axonal proliferation following lesions of cerebellar peduncles. A combined fluorescence microscopic and radioautographic study. J. comp. Neurol. **155**, 43–60 (1974b).

POSTERNAK, T., SUTHERLAND, E.W., HENION, W.F.: Derivatives of cyclic 3′5′-adenosine monophosphate. Biochim. biophys. Acta (Amst.) **65**, 558–560 (1962).

PRASAD, K.N., GILMER, K., KUMAR, S.: Morphologically "differentiated" mouse neuroblastoma cells induced by noncyclic AMP agents: levels of cyclic AMP, nucleic acid and protein. Proc. Soc. exp. Biol. (N.Y.) **143**, 1168–1171 (1973a).

PRASAD, K.N., HSIE, A.W.: Morphologic differentiation of mouse neuroblastoma cells induced in vitro by dibutyryl adenosine 3′:5′-cyclic monophosphate. Nature (Lond.) **233**, 141–142 (1971).

PRASAD, K.N., MANDAL, B., WAYMIRE, J.C., LEES, G.J., VERNADAKIS, A., WEINER, N.: Basal level of neurotransmitter synthesizing enzymes and effect of cyclic AMP agents on the morphological differentiation of isolated neuroblastoma clones. Nature (Lond.) **241**, 117–119 (1973b).

PURPURA, D.P., SHOFER, R.J.: Excitatory action of dibutyryl cyclic adenosine monophosphate on immature cerebral cortex. Brain Res. **38**, 179–181 (1972).

RALL, T.W.: Role of adenosine 3′,5′-monophosphate (cyclic AMP) in actions of catecholamines. Pharmacol. Rev. **24**, 399–409 (1972).

RALL, T.W., GILMAN, A.G.: The role of cyclic AMP in the nervous system. Neurosci. Res. Program Bull. **8**, (3) 221–317 (1970).

RALL, T.W., SATTIN, A.: Factors influencing the accumulation of cyclic AMP in brain tissue. Advanc. Biochem. Psychopharmacol. **3**, 113–133 (1970).

RALL, T.W., SUTHERLAND, E.W., BERTHET, J.: The relationship of epinephrine and glucagon to liver phosphorylase. IV. Effect of epinephrine and glucagon on the reactivation of phosphorylase in liver homogenates. J. biol. Chem. **224**, 463–475 (1957).

RALL, T.W., SUTHERLAND, E.: Formation of a cyclic adenine ribonucleotide by tissue particles. J. biol. Chem. **232**, 1065–1076 (1958).

RALL, T.W., SUTHERLAND, E.W.: The regulatory role of adenosine 3′,5′-phosphate. Cold Spr. Harb. Symp. quant. Biol. **26**, 347–354 (1961).

RALL, T.W., SUTHERLAND, E.W.: Adenyl cyclase. II. The enzymatically catalyzed formation of adenosine 3′-5′ phosphate and inorganic pyrophosphate from adenosine triphosphate. J. biol. Chem. **237**, 1228–1232 (1962).

RASMUSSEN, H.: Cell communication, calcium ion, and cyclic adenosine monophosphate. Science **170**, 404–412 (1970).

REIK, L., PETZOLD, G.L., HIGGINS, J.A., GREENGARD, P., BARRNETT, R.J.: Hormone-sensitive adenyl cyclase: cytochemical localization in rat liver. Science **168**, 382–384 (1970).

RICHELSON, E.: Stimulation of tyrosine hydroxylase activity in an adrenergic clone of mouse neuroblastoma by dibutyryl cyclic AMP. Nature (Lond.) **242**, 175–177 (1973).

ROBERTIS, E. DE, RODRIGUEZ DE LORES ARNAIZ, G., ALBERICI, M., BUTCHER, R.W., SUTHERLAND, E.W.: Subcellular distribution of adenyl cyclase and cyclic phosphodiesterase in rat brain cortex. J. biol. Chem. **242**, 3487–3493 (1967).

ROBISON, G.A., BUTCHER, R.W., SUTHERLAND, E.W.: Cyclic AMP. New York: Academic Press 1971.

RODBELL, M.: The role of nucleotides in the activity and response of adenylate cyclase to hormones. In: Advance in Cyclic Nucleotide Research, vol. 5. New York: Raven Press (in press).

RODBELL, M., BIRNBAUMER, L., POHL, S.L.: Hormones, receptors and adenyl cyclase activity in mammalian cells. In: The Role of Adenyl Cyclase and Cyclic 3′,5′-AMP in Biological Systems, Fogarty International Center Proceedings. Washington, D.C.: U.S. Government Printing Office 1971.

ROISEN, F.J., MURPHY, R.A.: Neurite development in vitro. II. The role of microfilaments and microtubules in dibutyryl adenosine 3′,5′-cyclic monophosphate and nerve-growth factor stimulated maturation. J. Neurobiol. **4**, 397–417 (1973).

ROISEN, F.J., MURPHY, R.A., BRADEN, W.G.: Dibutyryl cyclic adenosine monophate stimulation of colcemid-inhibited axonal elongation. Science **177**, 809–811 (1972a).

ROISEN, F.J., MURPHY, R.A., BRADEN, W.G.: Neurite development in vitro. I. The effects of adenosine 3′,5′-cyclic monophosphate (cyclic AMP). J. Neurobiol. **4**, 347–368 (1972b).

ROISEN, F.J., MURPHY, R.A., PICHICHERO, M.E., BRADEN, W.G.: Cyclic adenosine monophosphate stimulation of axonal elongation. Science **175**, 73–74 (1972c).

ROZEAR, M., DEGROOF, R., SOMJEN, G.: Effects of microiontophoretic administration of divalent metal ions on neurons of the central nervous system of cats. J. Pharmacol. exp. Ther. **176**, 109–118 (1971).

RUBIN, R.P., JAANUS, S.D., CARCHMAN, R.A.: Role of calcium and adenosine cyclic 3′-5′ phosphate in action of adreno corticotropin. Nature (Lond.) **240**, 150–152 (1972).

RUDLAND, P.S., GOSPODAROWICZ, D., SIEFERT, W.E.: Cyclic GMP and growth control in cultured fibroblasts: activation of guanyl cyclase and intracellular cGMP by a purified growth factor. Nature (Lond.) **250**, 741–742 (1974).

RUDOLPH, S.A., JOHNSON, E.M., GREENGARD, P.: The entholpy of hydrolysis of various 3′,5′- and 2′,3′-cyclic nucleotides. J. biol. Chem. **246**, 1271–1273 (1971).

RUSSELL, J.R., THOMPSON, W.J., SCHNEIDER, F.W., APPLEMAN, M.M.: 3′,5′-Cyclic adenosine monophosphate phosphodiesterase: negative cooperativity. Proc. nat. Acad. Sci. (Wash.) **69**, 1791–1795 (1972).

RUSSELL, T.R., PASTAN, I.H.: Cyclic adenosine 3′:5′-monophosphate and cyclic guanosine 3′:5′-monophosphate phosphodiesterase activities are under separate genetic control. J. biol. Chem. **249**, 7764–7769 (1974).

SAKAI, K., MARKS, B.: Adrenergic effects on pineal cell membrane potential. Life Sci. **11**, 285–291 (1972).

SALMOIRAGHI, G.C., BLOOM, F.E.: The pharmacology of individual neurons. Science **144**, 493–497 (1964).

SASA, M., MUNEKIYO, K., IKEDA, H., TAKAORI, S.: Noradrenaline-mediated inhibition by locus coeruleus of spinal trigeminal neurons. Brain Res. **80**, 443–460 (1974).

SATTIN, A., RALL, T.W.: The effect of adenosine and adenine nucleotides on the cyclic adenosine 3′,5′-phosphate content of guinea pig cerebral cortex slices. Molec. Pharmacol. **6**, 13–23 (1970).

SATTIN, A., RALL, T.W., ZANELLA, J.: Regulation of cyclic adenosine 3′,5′-monophosphate levels in guinea pig cerebral cortex by interaction of alpha adrenergic and adenosine receptor activity. J. Pharmacol. exp. Ther. **192**, 22–32 (1975).

SAYERS, G., BEALL, R.J., SEELIG, S.: Isolated adrenal cells: adreno-corticotropic hormone, calcium, steroidogenesis and cyclic adenosine monophosphate. Science **175**, 1131–1133 (1972).

SCHIMMER, B.P.: Effects of catecholamines and monovalent cations on adenylate cyclase activity in cultured glial tumor cells. Biochim. biophys. Acta (Amst.) **252**, 567–573 (1971).

SCHIMMER, B.P.: Influence of Li^+ on epinephrine-stimulated adenylate cyclase activity in cultured glial tumor cells. Biochim. biophys. Acta (Amst.) **326**, 186–192 (1973).

SCHMIDT, M.J., HOPKINS, J.T., SCHMIDT, D.E., ROBISON, G.A.: Cyclic AMP in brain areas: effects of amphetamine and norepinephrine assessed through the use of microwave radiation as a means of tissue fixation. Brain Res. **42**, 465–477 (1972).

SCHMIDT, M.J., PALMER, E.C., DETTBORN, W.-D., ROBISON, G.A.: Cyclic AMP and adenyl cyclase in the developing rat brain. Develop. Psychobiol. **3**, 53–67 (1970).

SCHMIDT, M.J., ROBISON, G.A.: The effect of norepinephrine on cyclic AMP levels in discrete regions of the developing rabbit brain. Life Sci. **10**, 459–464 (1971).

SCHMIDT, M.J., SCHMIDT, D.E., ROBISON, G.A.: Cyclic adenosine monophosphate in brain areas: microwave irradiation as a means of tissue fixation. Science **173**, 1142–1143 (1971).

SCHMIDT, M.J., SOKOLOFF, L.: Activity of cyclic AMP-dependent microsomal protein kinase and phosphorylation of ribosomal protein in rat brain during postnatal development. J. Neurochem. **21**, 1193–1205 (1973).

SCHROEDER, J.: Analogs of α-tocopherol as inhibitors of cyclic AMP and cyclic GMP phosphodiesterases and effects of α-tocopherol deficiency on cyclic AMP-controlled metabolism. Biochim. biophys. Acta (Amst.) **343**, 173–181 (1974).

SCHULTZ, G., BÖHME, E., MUNSKE, K.: Guanyl cyclase: Determination of enzyme activity. Life Sci. **8**, 1323–1332 (1969).

SCHULTZ, G., HARDMAN, J.G., SCHULTZ, K., BAIRD, C.E., SUTHERLAND, E.W.: The importance of calciums ions for the regulation of guanosine 3′,5′-cyclic monophosphate levels. Proc. nat. Acad. Sci. (Wash.) **70**, 3889–3893 (1973).

SCHULTZ, J.: Inhibition of phosphodiesterase activity in brain cortical slices from guinea pig and rat. Pharmacol. Res. Commun. **6**, 335–341 (1974).

SCHULTZ, J., DALY, J.W.: Cyclic adenosine 3′,5′-monophosphate in guinea pig cerebral cortical slices. II. The role of phosphodiesterase activity in the regulation of levels of cyclic adenosine 3′,5′-monophosphate. J. biol. Chem. **248**, 853–859 (1973a).

SCHULTZ, J., DALY, J.W.: Cyclic adenosine 3′,5′-monophosphate in guinea pig cerebral cortical slices. III. Formation, degradation, and reformation of cyclic adenosine 3′,5′-monophosphate during sequential stimulations by biogenic amines and adenosine. J. biol. Chem. **248**, 860–866 (1973b).

SCHULTZ, J., DALY, J.W.: Adenosine 3′,5′-monophosphate in guinea pig cerebral cortical slices: effects of α- and β-adrenergic agents, histamine, serotonin, and adenosine. J. Neurochem. **21**, 573–579 (1973c).

SCHULTZ, J., DALY, J.W.: Accumulation of cyclic adenosine 3′,5′-monophosphate in cerebral cortical slices from rat and mouse stimulatory effect of α- and β-adrenergic agents and adenosine. J. Neurochem. **21**, 1319–1326 (1973d).

SCHULTZ, J., HAMPRECHT, B., DALY, J.W.: Accumulation of adenosine 3′,5′-cyclic monophosphate in clonal glial cells: labeling of intracellular adenine nucleotides with radioactive adenine. Proc. nat. Acad. Sci. (Wash.) **69**, 1266–1270 (1972).

SCHWARTZ, J.P., MORRIS, N.R., BRECKENRIDGE, B.M.: Adenosine 3′,5′-monophosphate in glial tumor cells. J. biol. Chem. **248**, 2699–2704 (1973).

SEEDS, N.W., GILMAN, A.G.: Norepinephrine stimulated increase of cyclic AMP levels in developing mouse brain cell cultures. Science **174**, 292 (1971).

SEEMAN, P., LEE, T.: Tranquilizer-blockade of dopamine release from stimulated striatal slices. 4th Annu. Meeting Soc. for Neurosci., St. Louis, 620, 1974.

SEGAL, M., BLOOM, F.E.: The action of norepinephrine in the rat hippocampus. I. Iontophoretic studies. Brain Res. **72**, 79–97 (1974a).

SEGAL, M., BLOOM, F.E.: The action of norepinephrine in the rat hippocampus. II. Activation of the input pathway. Brain Res. **72**, 99–114 (1974b).

SHASHOUA, V.E.: Dibutyryl adenosine cyclic 3′,5′-monophosphate effects on goldfish behavior and brain RNA metabolism. Proc. nat. Acad. Sci. (Wash.) **68**, 2835–2838 (1971).

SHEIN, H.M., WURTMAN, R.J.: Cyclic adenosine monophosphate: stimulation of melatonin and serotonin synthesis in cultured rat pineals. Science **166**, 519–520 (1969).

SHEPPARD, H., BURGHARDT, C.R.: The dopamine-sensitive adenylate cyclase of rat caudate nucleus. I. Comparison with the isoproterenol-sensitive adenylate cyclase (beta receptor system) of rat erythrocytes in responses to dopamine derivatives. Molec. Pharmacol. **10**, 721–726 (1974).

SHEPPARD, J.R., HUDSON, T.H., LARSON, J.R.: Adenosine 3′,5′-monophosphate analogus promote a circular morphology of cultured schwannoma cells. Science **187**, 179–181 (1975).

SHEPPARD, J.R., PRASAD, K.N.: Cyclic AMP levels and the morphological differentiation of mouse neuroblastoma cells. Life Sci. **12**, 431–439 (1973).

SHERRINGTON, C.S.: Remarks on some aspects of reflex inhibition. Proc. roy. Soc. B **97**, 519–544 (1925).

SHIMIZU, H., DALY, J.: Formation of cyclic adenosine 3′,5′ monophosphate from adenosine in brain slices. Biochim. biophys. Acta (Amst.) **222**, 465–473 (1970).

SHIMIZU, H., DALY, J.W.: Effect of depolarizing agents on accumulation of cyclic adenosine 3′,5′-monophosphate in cerebral cortical slices. Europ. J. Pharmacol. **17**, 240–252 (1972).

SHIMIZU, H., CREVELING, C.R., DALY, J.: Stimulated formation of adenosine 3′,5′-cyclic phosphate in cerebral cortex: synergism between electrical activity and biogenic amines. Proc. nat. Acad. Sci. (Wash.) **65**, 1033–1040 (1970a).

SHIMIZU, H., CREVELING, C.R., DALY, J.W.: Cyclic adenosine 3′,5′-monophosphate formation in brain slices: stimulation by batrachotoxin, ouabain, veratridine and potassium ions. Molec. Pharmacol. **6**, 184–188 (1970b).

SHIMIZU, H., CREVELING, C.R., DALY, J.W.: The effect of histamines and other compounds on the formation of adenosine 3′,5′-monophosphate in slices from cerebral cortex. J. Neurochem. **17**, 441–444 (1970c).

SHIMIZU, H., ICHISITA, H., ODAGIRI, H.: Stimulated formation of cyclic adenosine 3′,5′-monophosphate by aspartate and glutamate in cerebral cortical slices of guinea pig. J. biol. Chem. **249**, 5955–5962 (1974).

SHIMIZU, H., TANAKA, S., SUZUKI, T., MATSUKADO, Y.: The response of human cerebrum adenyl cyclase to biogenic amines. J. Neurochem. **18**, 1157–1161 (1971).

SHOEMAKER, W.J., BALENTINE, L.T., SIGGINS, G.R., HOFFER, B.J., HENRIKSEN, S.J., BLOOM, F.E.: Characteristics of the release of cyclic adenosine 3′,5′-monophosphate from micropipets by microiontophoresis. J. Cyclic Nucleotide Res. **1**, 97–106 (1975).

SIGGINS, G.R., BATTENBERG, E.F., HOFFER, B.J., BLOOM, F.E., STEINER, A.L.: Noradrenergic stimulation of cyclic adenosine monophosphate in rat Purkinje neurons: an immunocytochemical study. Science **179**, 585–588 (1973).

SIGGINS, G.R., HENRIKSEN, S.J.: Inhibition of rat Purkinje neurons by analogues of cyclic adenosine monophosphate: correlation with protein kinase activation. Science **189**, 557–560 (1975).

SIGGINS, G.R., HOFFER, B.J., BLOOM, F.E.: Cyclic 3′,5′-adenosine monophosphate: possible mediator for the response of cerebellar Purkinje cells to microelectrophoresis of norepinephrine. Science **165**, 1018–1020 (1969).

SIGGINS, G.R., HOFFER, B.J., BLOOM, F.E.: Studies on norepinephrine-containing afferents to Purkinje cells of rat cerebellum. III. Evidence for mediation of norepinephrine effects by cyclic 3′,5′-adenosine monophosphate. Brain Res. **25**, 535–553 (1971a).

SIGGINS, G.R., HOFFER, B.J., OLIVER, A.P., BLOOM, F.E.: Activation of a central noradrenergic projection to cerebellum. Nature (Lond.) **233**, 481–483 (1971b).

SIGGINS, G.R., HOFFER, B.J., BLOOM, F.E.: Prostaglandin-norepinephrine interactions in brain: microelectrophoretic and histochemical correlates. Ann. N.Y. Acad. Sci. **180**, 302–323 (1971c).

SIGGINS, G.R., HOFFER, B.J., UNGERSTEDT, U.: Electrophysiological evidence for involvement of cyclic adenosine monophosphate in dopamine responses of caudate neurons. Life Sci. **15**, 779–792 (1974).

SIGGINS, G.R., OLIVER, A.P., HOFFER, B.J., BLOOM, F.E.: Cyclic adenosine monophosphate and norepinephrine: effects on transmembrane properties of cerebellar Purkinje cells. Science **171**, 192–194 (1971d).

SIMON, L.N., SHUMAN, D.A., ROBINS, R.K.: The synthesis and biological activity of analogs of cyclic nucleotides. In: Cellular Mechanisms. Basel: S. Karger 1973.

SKOLNICK, P., DALY, J.W.: The accumulation of adenosine 3′,5′-monophosphate in cerebral cortical slices of the awaking mouse, a neurologic mutant. Brain Res. **73**, 513–525 (1974).

SKOLNICK, P., HUANG, M., DALY, J., HOFFER, B.J.: Accumulation of adenosine 3′,5′-monophosphate in incubated slices from discrete regions of squirrel monkey cerebral cortex: effect of norepinephrine, serotonin, and adenosine. J. Neurochem. **21**, 237–240 (1973).

SLOBODA, R.D., RUDOLPH, S.A., ROSENBAUM, J.L., GREENGARD, P.: Cyclic AMP-dependent endogenous phosphorylation of a microtubule-associated protein. Proc. nat. Acad. Sci. (Wash.) **72**, 177–181 (1975).

SOIFER, D.: Enzymatic activity in tubulin preparations: Cyclic AMP dependent protein kinase activity of brain microtubule protein. J. Neurochem. **24**, 21–33 (1975).

SOMLYO, A.V., HAEUSLER, G., SOMLYO, A.P.: Cyclic adenosine monophosphate: potassium-dependent action on vascular smooth muscle membrane potential. Science **169**, 490–491 (1970).

SPIEGEL, A.M., AURBACH, G.D.: Binding of 5′-guanylyl-imidodiphosphate to turkey erythrocyte membranes and effects on β-adrenergic activated adenylate cyclase. J. biol. Chem. **249**, 7630–7636 (1974).

STAVINOHA, W.B., PEPELKO, B., SMITH, P.: Microwave radiation to inactivate cholinesterase in the rat brain prior to analysis for acetylcholine. Pharmacologist **12**, 257 (1970).

STEINER, A.L., PARKER, C.W., KIPNIS, D.M.: The measurement of cyclic nucleotides by radioimmunoassay. Advanc. Biochem. Psychopharmacol. **3**, 89–112 (1970).

STEINER, A.L., PARKER, C.W., KIPNIS, D.M.: Radioimmuno-assay for cyclic nucleotides. I. Preparation of antibodies and iodinated cyclic nucleotides. J. biol. Chem. **247**, 1106–1113 (1972a).

STEINER, A.L., PAGHARA, A.W., CHASE, L.R., KIPNIS, D.M.: Radioimmunoassay for cyclic nucleodies. II. Adenosine 3′,5′-monophosphate and guanosine 3′,5′-monophosphate in mammalian tissues and body fluids. J. biol. Chem. **247**, 1114–1120 (1972b).

STEINER, A.L., FERRENDELLI, J.A., KIPNIS, D.M.: Radioimmunoassay for cyclic nucleotides. III. Effect of ischemia, changes during development and regional distribution of adenosine 3′,5′-monophosphate and guanosine 3′,5′ monophosphate in mouse brain. J. biol. Chem. **247**, 1121–1124 (1972c).

STONE, T.W., TAYLOR, D.A., BLOOM, F.E.: Cyclic AMP and cyclic GMP may mediate opposite neuronal responses in the rat cerebral cortex. Science **187**, 845–847 (1975).

STRADA, S.J., UZUNOV, P., AND WEISS, B.: Ontogenetic development of a phosphodiesterase activator and the multiple forms of cyclic AMP phosphodiesterase of rat brain. J. Neurochem. **23**, 1097–1103 (1974).

SUTHERLAND, E.W., OYE, I., BUTCHER, R.W.: The action of epinephrine and the role of the adenyl cyclase system in hormone action. Recent Progr. Hormone Res. **21**, 623–642 (1965).

SUTHERLAND, E.W., RALL, T.W.: Fractionation and characterization of a cyclic adenine ribonucleotide formed by tissue particles. J. biol. Chem. **232**, 1077–1091 (1958).

SUTHERLAND, E.W., RALL, T.W., MENON, T.: Adenyl cyclase. I. Distribution, preparation and properties. J. biol. Chem. **237**, 1220–1227 (1962).

SWILLENS, S., CAUTER, E. VON, DUMONT, J.E.: Protein kinase and cyclic 3′,5′-AMP: significance of binding and activation constants. Biochim. biophys. Acta (Amst.) **364**, 250–259 (1974).

TADA, M., KIRCHBERGER, M.A., REPKE, D.I., KATZ, A.M.: The stimulation of calcium transport in cardiac sarcoplasmic reticulum by adenosine 3′,5′-monophosphate dependent protein kinase. J. biol. Chem. **249**, 6174–6180 (1974).

TAGLIAMONTE, A., TAGLIAMONTE, P., FORN, J., PEREZ-CRUET, J., KRISHNA, G., GESSA, G.L.: Stimulation of brain serotonin synthesis by dibutyryl cyclic AMP in rats. J. Neurochem. **18**, 1191–1196 (1971).

TAKAMORI, M., ISHII, N., MORI, M.: The role of cyclic 3′,5′-adenosine monophosphate in neuromuscular transmission. Arch. Neurol. **29**, 420–424 (1973).

TAMARIND, D.C., QUILLIAM, J.P.: Synaptic organization and other ultrastructural features of the superior cervical ganglion of the rat, kitten and rabbit. Micron **2**, 204–234 (1971).

TESHIMA, Y., YAMAZAKI, R., KAKIUCHI, S.: Effects of ATP on the activity of nucleoside 3′,5′-cyclic monophosphate phosphodiesterase from brain. J. Neurochem. **22**, 789–791 (1974).

THIERRY, A.M., BLANC, G., SOBEL, A., STINUS, L., GLOWINSKI, J.: Dopaminergic terminals in the rat cortex. Science **182**, 499–501 (1973).

THOENEN, H.: Induction of tyrosine hydroxylase in peripheral and central adrenergic neurones by cold exposure of rats. Nature (Lond.) **228**, 861–862 (1970).

THOENEN, H., MUELLER, R.A., AXELROD, J.: Trans-synaptic induction of adrenal tyrosine hydroxylase. J. Pharmacol. exp. Ther. **169**, 249–254 (1969).

THOENEN, H., OTTEN, U., OESCH, F.: Trans-synaptic regulation of tyrosine hydroxylase. In: Frontiers in Catecholamine Research. New York: Pergamon Press 1974.

THOMPSON, W.J., APPLEMAN, M.M.: Multiple cyclic nucleotide phosphodiesterases activities from rat brain. Biochemistry (Wash.) **10**, 311–316 (1971).

TOMITA, T., SAKAMOTO, Y., OHBA, M.: Conductance increase by adrenaline in guinea pig taenia coli studied with voltage clamp method. Nature (Lond.) **250**, 432 (1974).

TRIGGLE, D.J.: Neurotransmitter-receptor Interactions. New York: Academic Press 1971.

TSIEN, R.W.: Adrenaline-like effects of intracellular iontophoresis of cyclic AMP in cardiac Purkinje fibres. Nature (Lond.) **245**, 120–122 (1973).

TSIEN, R.W., GILES, W., GREENGARD, P.: Cyclic AMP mediates the effects of adrenaline on cardiac Purkinje fibers. Nature (Lond.) **240**, 181–183 (1972).

TSIEN, R.W., WEINGART, R.: Cyclic AMP: Cell-to-cell movement and inotropic effect in ventricular muscle, studied by a cut-end method. J. Physiol. (Lond.) **242**, 95–96 (1974).

TSOU, K.C., YIP, K.F., LO, K.W.: 1,N^6 Etheno-2-aza-adenosine 3′,5′-monophosphate: a new fluorescent substrate for cycle nucleotide phosphodiesterase. Analyt. Biochem. **60**, 163–169 (1974).

UEDA, T., MAENO, H., GREENGARD, P.: Regulation of endogenous phosphorylation of specific proteins in synaptic membrane fractions from rat brain by adenosine 3′5′-monophosphate. J. biol. Chem. **248**, 8295–8305 (1973).

UNGERSTEDT, U.: Stereotaxic mapping of the monoamine pathways in rat brain. Acta physiol. scand. **67**, 1–48 (1971).

UZUNOV, P., REVUELTA, A., COSTA, E.: A role for the endogenous activator of 3′,5′-nucleotide phosphodiesterase in rat adrenal medulla. Molec. Pharmacol. (in press, 1975).

UZUNOV, P., SHEIN, H.M., WEISS, B.: Cyclic AMP phosphodiesterase in cloned astrocytoma cells: norepinephrine induces a specific enzyme form. Science **180**, 304–306 (1973).

UZUNOV, P., SHEIN, H.M., WEISS, B.: Multiple forms of cyclic 3′,5′-AMP phosphodiesterase of rat cerebrum and cloned astrocytoma and neuroblastoma cells. Neuropharmacol. **13**, 377–391 (1974).

UZUNOV, P., WEISS, B.: Effects of phenothiazine tranquillizers on the cyclic 3′,5′-adenosine monophosphate system of rat brain. Neuropharmacol. **10**, 697–708 (1971).

UZUNOV, P., WEISS, B.: Psychopharmacological Agents and the cyclic AMP system of rat brain. In: Advances in Cyclic Nucleotide Research, vol. 1. New York: Raven Press 1972.

VASSALLE, M., BARNABEL, O.: Norepinephrine and potassium fluxes in cardiac Purkinje fibers. Pflügers Arch. **322**, 287–303 (1971).

VEECH, R.L., HARRIS, R.L., VELOSO, D., VEECH, E.H.: Freezeblowing: a new technique for the study of brain in vivo. J. Neurochem. **20**, 183–188 (1973).

VELLIS, J. DE, BROOKER, G.: Reversal of catecholamine refractoriness by inhibitors of RNA and protein synthesis. Science **186**, 1221–1222 (1974).

VERMA, S.C., MCNEILL, J.H.: Action of imidazole on the cardiac inotropic, phosphorylase activating and cyclic AMP producing effects of norepinephrine and histamine. Res. Commun. Chem. Pathol. Pharmacol. **7**, 305–319 (1974).

VESIN, M.F., HARBON, S.: The effect of epinephrine, prostaglandins, and their antagonists on adenosine cyclic 3′,5′-monophosphate concentrations and motility of the rat uterus. Molec. Pharmacol. **10**, 457–473 (1974).

VOADEN, M.J.: The effects of superior cervical ganglionectomy and/or bilateral adrenalectomy on the mitotic acitivity of the adult rat cornea. Exp. Eye Res. **12**, 337–341 (1971).

VOKAER, A., IACOBELLI, S., KRAM, R.: Phosphoprotein phosphatase activity associated with estrogen-induced protein in rat uterus. Proc. nat. Acad. Sci. (Wash.) **71**, 4482–4486 (1974).

VOLICER, L., GOLD, B.I.: Effect of ethanol on cyclic AMP levels in the rat brain. Life Sci. **13**, 269–280 (1973).
VON HUNGEN, K., ROBERTS, S.: Catecholamine and Ca^{2+} activation of adenylate cyclase systems in synaptosomal fractions from rat cerebral cortex. Nature (Lond.) **242**, 58–60 (1973).
VON HUNGEN, K., ROBERTS, S.: Neurotransmitter-sensitive adenylate cyclase systems in the brain. In: Reviews of Neuroscience, vol. 1. New York: Raven Press 1974.
WAGNER, R.C., BITENSKY, M.W.: Adenylate Cyclase. In: Electron Microscopy of Enzymes. New York: Van Nostrand, Rienhold Co. 1974.
WAGNER, R.C., KRIENER, P., BARRNETT, R.J., BITENSKY, M.W.: Biochemical characterization and cytochemical localization of a catecholamine-sensitive adenylate cyclase in isolated capillary endothelium. Proc. nat. Acad. Sci. (Wash.) **69**, 3175–3180 (1972).
WALKER, J.B., WALKER, J.P.: Neurohumoral regulation of adenylate cyclase activity in rat striatum. Brain Res. **54**, 386–390 (1973a).
WALKER, J.B., WALKER, J.P.: Properties of adenylate cyclase from senescent rat brain. Brain Res. **54**, 391–396 (1973b).
WALSH, D.A., PERKINS, J.P., KREBS, E.G.: An adenosine 3′,5′ monophosphate dependent protein kinase from rabbit skeletal muscle. J. biol. Chem. **243**, 3763–3765 (1968).
WATANABE, A.M., BESCH, H.R. JR.: Cyclic adenosine monophosphate modulation of slow calcium influx channels in guinea pig heart. Circulat. Res. **35**, 316–324 (1974).
WAYMIRE, J.C., WEINER, N., PRASAD, K.N.: Regulation of tyrosine hydroxylase activity in cultured mouse neuroblastoma cells: elevation induced by analogs of adenosine 3′,5′ cyclic monophosphate. Proc. nat. Acad. Sci. (Wash.) **69**, 2241–2245 (1972).
WEBER, A., MURRAY, J.M.: Molecular control mechanisms in muscle contraction. Physiol. Rev. **53**, 612–673 (1973).
WEDNER, H.J., HOFFER, B.J., BATTENBERG, E., STEINER, A.L., PARKER, C.W., BLOOM, F.E.: A method for detecting intracellular cyclic adenosine monophosphate by immunofluorescence. J. Histochem. Cytochem. **20**, 293–295 (1972).
WEIGHT, F.F.: Mechanisms of synaptic transmission. Neurosci. Res. Program Bull. **4**, 1–27 (1971).
WEIGHT, F.F.: Physiological mechanisms of synaptic modulation. In: The Neurosciences: 3rd Study Program. Cambridge: MIT Press 1974.
WEIGHT, F.F., PADJEN, A.: Acetylcholine and slow synaptic inhibition in frog sympathetic ganglion cells. Brain Res. **55**, 225–228 (1973a).
WEIGHT, F.F., PADJEN, A.: Slow synaptic inhibition: evidence for synaptic inactivation of sodium conductance in sympathetic ganglion cells. Brain Res. **55**, 219–224 (1973b).
WEIGHT, F.F., PETZOLD, G., GREENGARD, P.: Guanosine 3′,5′-monophosphate in sympathetic ganglia: increase associated with synaptic transmission. Science **186**, 942–944 (1974).
WEIGHT, F.F., VOTAVA, J.: Slow synaptic excitation in sympathetic ganglion cells: evidence for synaptic activation of potassium conductance. Science **170**, 755–758 (1970).
WEINRUB, I., CHASIN, M., FREE, C.A., HARRIS, D.N., GOLDENBERG, H., MICHEL, I.M., PALK, U.S., PHILLIPS, M., SAMANIEGO, S., HESS, S.M.J.: Effects of therapeutic agents on cyclic AMP metabolism in vitro. J. pharm. Sci. **61**, 1556–1657 (1972).
WEISS, B.: Psychopharmacological agents and the cyclic AMP system of rat brain. In: Advances in Cyclic Nucleotide Research, vol. 1. New York: Raven Press 1972.
WEISS, B., COSTA, E.: Regional and subcellular distribution of adenyl cyclase and 3′,5′-cyclic nucleotide phosphodiesterase in brain and pineal gland. Biochem. Pharmacol. **17**, 2107–2116 (1968).
WEISS, B., KIDMAN, A.D.: Neurobiological significance of cyclic 3′,5′-adenosine monophosphate. Advanc. Biochem. Psychopharmacol. **1**, 132–164 (1969).
WELLER, M., RODNIGHT, R.: Stimulation by cyclic AMP of intrinsic protein kinase activity in ox brain membrane preparations. Nature (Lond.) **225**, 187–188 (1970).
WERMAN, R.: CNS cellular level: membranes. Ann. Rev. Physiol. **34**, 337–374 (1972).
WHITE, A.A., AURBACH, G.D.: Detection of guanyl cyclase in mammalian tissues. Biochim. biophys. Acta (Amst.) **191**, 686–697 (1969).
WHITFIELD, J.F., RIXON, R.H., MCMANUS, J.P., BALK, S.D.: Calcium, cyclic adenosine 3′,5′-monophosphate, and the control of cell proliferation: a review In Vitro. **8**, 257–278 (1973).
WICKS, W.D.: Regulation of protein synthesis by cyclic AMP. In: Advances in Cyclic Nucleotide Research, vol. 4. New York: Raven Press 1974.
WICKSON, R.D., BOUDREAU, R.J., DRUMMOND, G.I.: Activation of 3′,5′-cyclic adenosine monophos-

phate phosphodiesterase by calcium ion and a protein activator. Biochemistry (Wash.) **14**, 669–675 (1975).

WILL, H., SCHIRPKE, B., WOLLENBERGER, A.: Binding of calcium to a cell membrane-enriched preparation from pig myocardium: increase in calcium affinity upon membrane protein phosphorylation enhanced by a membrane-bound cyclic AMP dependent protein kinase. Acta biol. med. germ. **31**, 45–52 (1973).

WILLIAMS, B.J., PIRCH, J.H.: Correlation between brain adenyl cyclase activity and spontaneous motor activity in rats after chronic reserpine treatment. Brain Res. **68**, 227–234 (1974).

WILLIAMS, M., RODNIGHT, R.: Evidence for a role for protein phosphorylation in synaptic function in the cerebral cortex mediated through a β-noradrenergic receptor. Brain Res. **77**, 502–506 (1974).

WILLIAMS, R.H., THOMPSON, W.J.: Effect of age upon guanyl cyclase, adenyl cyclase, and cyclic nucleotide phosphodiesterases in rats. Proc. Soc. exp. Biol. (N.Y.) **143**, 382–387 (1973).

WILLIAMS, T.H., PALAY, S.L.: Ultrastructure of the small neurons in the superior cervical ganglion. Brain Res. **15**, 17–34 (1969).

WILSON, D.F.: The effects of dibutyryl cyclic adenosine 3′,5′-monophosphate, theophylline, and aminophylline on neuromuscular transmission in the rat. J. Pharmacol. exp. Ther. **188**, 447–452 (1974).

WILSON, D.F., STUBBS, M., VEECH, R.L., ERECINSKA, M., KREBS, H.A.: Equilibrium relations between the oxidation-reduction reactions and the adenosine triphosphate synthesis in suspensions of isolated liver cells. Biochem. J. **140**, 57–64 (1974).

WOODWARD, D.J., HOFFER, B.J., ALTMAN, J.: Physiological and pharmacological properties of Purkinje cells in rat cerebellum degranulated by postnatal X-irradiation. J. Neurobiol. **5**, 283–304 (1974).

YARBROUGH, G.G., LAKE, N., PHILLIS, J.W.: The role of calcium in monoamine induced depression of cerebral cortical neurones. Life Sci. **13**, 703–711 (1973).

YOKOTA, R.: The granule-containing cell somata in the superior cervical ganglion of the rat, as studied by a serial sampling method for electron microscopy. Z. Zellforsch. **141**, 331–346 (1973).

YORK, D.H.: Dopamine receptor blockade: a central action of chlorpromazine on striatal neurones. Brain Res. **37**, 91–99 (1972).

YOUNT, R.G., BOBCOCK, D., BALLENTYNE, W., OHALA, D.: Adenylyl-immidodiphosphate; an adenosine triphosphate analog containing a P-N-P linkage. Biochemistry (Wash.) **10**, 2484 (1971).

ZANELLA, J. JR., RALL, T.W.: Evaluation of electrical pulses and elevated levels of potassium ions as stimulants of adenosine 3′,5′-monophosphate (cyclic AMP) accumulation in guinea pig brain. J. Pharmacol. exp. Ther. **186**, 241–252 (1973).

ZIGMOND, R.E., SCHON, F., IVERSEN, L.L.: Increased tyrosine hydroxylase activity in the locus coeruleus of rat brain stem after reserpine treatment and cold stress. Brain Res. **70**, 547–552 (1974).

Rev. Physiol. Biochem. Pharmacol., Vol. 74

Renal Transport of Amino Acids

S. SILBERNAGL*, E.C. FOULKES**, and P. DEETJEN*

Contents

I. Introduction

The class of naturally occurring amino acids comprises over 20 compounds, mostly possessing an L-α-amino configuration. In circulating plasma they are present in free form in small but significant amounts, and as such are readily filtered at the glomerulus. Nevertheless, the urine concentration of amino acids is normally very low, due to efficient reabsorptive mechanisms localized mostly in proximal tubules.

Under certain abnormal conditions the concentration of free amino acids in urine is greatly increased, a phenomenon appropriately described as aminoaciduria. More than 150 years ago WOLLASTON [364] had already investigated the chemical properties of a stone which had crystallized out of urine in the bladder and discovered in it a sulfur-containing substance which he named cystine, from the Greek word for bladder. As more exact methods of measurements for amino acids became available aminoacidurias were reported with increasing frequency. Today many examples of such excretory malfunction are known. Thus, generalized tubular damage is likely to lead to increased excretion of all normally reabsorbed amino acids; in other cases specific lesions affect transport of only specific compounds.

* Physiologisches Institut der Universität Innsbruck, Fritz-Pregl-Str. 3, A-6020 Innsbruck/Austria.
** Depts. of Environmental Health and Physiology, Univ. of Cincinnati, Coll. of Medicine, Cincinnati, Ohio 45267, USA.

It is important to distinguish between two kinds of aminoaciduria. In the first variety increased amino acid excretion follows a pathological increase of the plasma amino acid concentration. This increase leads to an excessive filtered load which may exceed the reabsorptive capacity of the tubule, and thus cause a prerenal or overflow aminoaciduria. Included in this category are cases where an amino acid circulating in excessive amounts in plasma interferes with the renal reabsorption also of other amino acids. Excretion of these other amino acids may then greatly increase even though they are present at normal plasma concentrations.

The second variety of aminoaciduria represents an interference with tubular reabsorption mechanisms. Even at normal amino acid concentrations in plasma this causes an increased urinary excretion of amino acids and is therefore refered to as renal aminoaciduria. The pathogenesis and etiology of renal aminoaciduria are essentially unknown and little progress is to be expected in their clarification until the mechanism of tubular amino acid transport is more thoroughly understood. It is the purpose of this review to summarize critically progress in this area. Special emphasis will be laid on the contributions and limitations of the various experimental approaches which have provided the basis for our present knowledge. Hopefully we will succeed in throwing somewhat more light on the physiology of renal amino acid transport. This would contribute to an understanding not only of the pathogenesis of aminoaciduria but also of fundamental aspects of membrane transport.

II. Classification of Naturally Occuring Amino Acids

Traditionally the naturally occurring α-amino acids are separated on the basis of their chemical properties into three specific groups: (a) The neutral amino acids (monoaminomonocarboxylic or diaminodicarboxylic acids), (b) the cationic or basic amino acids, i.e. diaminomonocarboxylic acids, and (c) the anionic or acidic amino acids, i.e. monoaminodicarboxylic acids.

This simple and useful classification led to the expectation that three distinct systems would be found to account for the renal tubular transport of all amino acids. It was soon realized, however, that at least in the case of neutral amino acids no single reabsorption mechanism can explain the observed results. Indeed a final assignment of these neutral amino acids to particular transport systems is not yet possible.

The following section headings are therefore used without necessarily implying the existence of corresponding specific reabsorption systems:

1. Neutral amino acids
   a) glycine
   b) proline and hydroxyproline
   c) cystine and cysteine
   d) other neutral amino acids

2. Anionic amino acids such as glutamic and aspartic acids.
3. Cationic amino acids such as arginine, lysine and ornithine.

The usefulness of this subdivision of amino acids for the purpose of discussing their renal transport will become apparent in this review.

III. Evaluation of Experimental Techniques

Considerable knowledge has been collected on transport of various classes of amino acids in kidneys (see also recent reviews: [141, 142, 372]). Significant contributions to this understanding were first made by classical clearance determinations under various conditions, although analytical difficulties tended at times to obscure the results. Stop flow procedures helped localize a major portion of amino acid reabsorption in the proximal tubule [43, 136, 170, 267]. Subsequently ROSENBERG et al. [255a] adapted the slice technique of CROSS and TAGGART [78] to this field. Later advances were based mainly on the use of micropuncture and microperfusion techniques.

Each of these approaches can illuminate one or more aspects of the renal transport of amino acids. Thus, disappearance of an amino acid from a microperfused tubule clearly reflects movement of this acid in the direction of reabsorption. Similarly, an amino acid clearance smaller than the glomerular filtration rate provides a direct measure of reabsorption unless one supposes that simultaneous secretion of a compound also took place.

Use of *in vitro* techniques possesses the great advantage of precise control over composition of extracellular fluid, and of permitting extension of transport studies to unphysiological values of temperature, pH, electrolyte concentration etc.

The major limitation of renal slices arises from the fact that solute movement in such a system appears to reflect primarily events at the peritubular cell membrane. Slice data therefore shed little light on the process of amino acid reabsorption [115]. Indeed, questions have repeatedly been raised about the significance of such data. Thus problems may arise from uneven penetration of solutes into slices, and especially into inner portions of even relatively thin preparations [352]. More specifically doubts exist about participation of luminal cell membranes in solute exchange between slices and bathing medium, and therefore about the relevance of slice data to the elucidation of transport processes at the luminal membrane. Thus, FOULKES and MILLER [112] concluded that PAH efflux from rabbit kidney cortex slices only involved the peritubular membrane. This conclusion was further elaborated by MURTHY and FOULKES [231], in part on the basis that collapse of tubular lumina in slices prevents access from medium to luminal membranes [35].

More direct evidence that solute uptake by slices represents primarily transport across peritubular cell membranes was obtained by comparing movement of sugars and amino acids *in vivo* and *in vitro* [115]. In these experiments the double indicator dilution technique of SILVERMAN et al. [315] was used in intact animals to demonstrate peritubular transport. Not only was a striking similarity observed between peritubular transport and transport in slices but, in the case of aspartate, under certain conditions a clear distinction could be achieved between peritubular and slice transport on the one hand and reabsorptive movement on the other hand. A dissociation of reabsorption of L-lysine from its accumulation in rat kidney cortex was also reported by AUSIELLO et al. [11].

The restricted view here put forward of the usefulness of renal cortical slices has not been entirely accepted. One question arises from the kinetic analysis of amino acid accumulation by ROSENBERG et al. [255a]. These authors observed that cellular accumulation of amino acids in slices proceeds at a faster rate than that of their distribution in the inulin space. It was therefore concluded [285] that diffusion of amino acid through the extracellular space to the peritubular membrane could not form part of the process of cellular amino acid accumulation in slices. As a possible alternative, transport in the direction of reabsorption from lumen to cell was considered for the mechanism of this accumulation.

Now it may safely be assumed that if tubular lumina are accessible to diffusion of amino acids from the ambient medium then inulin must also reach these areas. Tubular lumina in other words would form part of the inulin space and the kinetic findings could be rephrased by stating that only a small portion of the total inulin space of the tissue lies on the diffusion path from medium to cells. Of interest here is the identical finding made in a study of a penetration of short-chain fatty acids into rat diaphragm [114].

In this case exclusion of the bulk of the extracellular fluid space from the process of solute uptake by cells can of course be attributed to asymmetric cells in special structures such as renal tubules. The kinetics of solute uptake by renal slices are thus fully compatible with the view that it is primarily the peritubular membrane which is involved in this process. The view that amino acid uptake by slices of renal cortex does not reflect reabsorptive processes is further strengthened by earlier observations that renal medullary and glomerular slices [207, 209], as well as slices of cerebral cortex [326] all possess the ability to concentrate amino acids. We may conclude that the peritubular transport systems studied in slices belong to the same class of pumps as those studied in Ehrlich ascites tumor cells [52] and other tissues.

In spite of the likelihood that tubular reabsorption of amino acids is not a process which can be studied to advantage in kidney slices, good correlations have repeatedly been established between properties of reabsorptive processes *in vivo* and accumulation of amino acids in slices. Many instances might be quoted. Thus, L-lysine and L-ornithine inhibit both reabsorption of L-arginine from the tubular lumen [22] and its accumulation by slices [256]. Similarly, L-lysine and L-arginine inhibition of L-cystine transport has been demonstrated both *in vitro* [361] and in micropuncture studies of amino acid reabsorption from the rat proximal tubule *in vivo* [301]. It is precisely these correlations which appeared to justify use of the slice technique and made it such a useful stimulus for complementary work by other approaches.

Instances have also been reported however where the correlation between reabsorption and *in vitro* transport is not complete. Thus, competition between accumulation of arginine and histidine in slices has been reported [256]; no such interaction was seen in studies of arginine reabsorption from the microperfused tubule [313], or during measurements of arginine clearance in the intact dog [138].

From the point of view of their respective functions, probable differences between the transport events at luminal and peritubular membranes are also clearly apparent. Thus, luminal mechanisms are primarily involved in amino acid reabsorption from the tubular lumen. This conclusion is derived both indirectly from measurement of blood-to-urine transit times of various amino acids in presence and absence of inhibitors of their reabsorption [118, 120], and also from direct visualization of amino acid transport by radioautography [242]. Peritubular transport on the other hand probably plays a primarily nutritive role, as further discussed in Section V. As the process of reabsorption is limited by events at the luminal membrane [120], direct measurement of amino acid disappearance from the lumen is clearly the most unequivocal approach to the study of amino acid reabsorption. Methods of choice for this purpose are therefore micropuncture of the tubule and allied techniques.

Use of micropuncture procedures goes back to 1937 when RICHARDS and WALKER [253] succeeded with fine glass capillaries in obtaining samples of tubular fluid from kidneys of cold-blooded animals. Analysis of such small samples proved extremely difficult however. Only the advances made in this area in later years permitted useful application of micropuncture techniques and made them an essential instrument of modern nephrological investigation (see review of the literature, [362]). Subsequently methods were also worked out permitting introduction of test solutions into single tubules [143, 292, 318]. SONNENBERG and DEETJEN [319] finally succeeded by means of a micropump in perfusing a single tubule at continuous and reproducible rates.

With certain modifications [86] such methods formed the basis of the microperfusion studies of amino acid reabsorption carried out by SILBERNAGL and DEETJEN [293, 295, 296, 300, 301, 302, 303] and SILBERNAGL [294, 297, 298, 299, 304, to 311]. In addition to tubular lumina, peritubular capillaries were also perfused [264, 293, 295, 321, 322]. Similar methods were employed in studies of amino acid transport by CHAN and HUANG [49] and CHRISTOPHEL and collaborators [58, 59, 60].

Continuous microperfusion [14, 86, 309, 319] permits perfusion of a single superficial tubular portion *in vivo* at a desired rate, as close as desired to physiological free flow conditions. In addition, for the proximal tubules the composition of a perfusion solution can be chosen in such a manner [185] that essentially no volume changes across the tubular epithelium will occur (steady state solution). The flow rate along the tubule therefore remains constant. This fact facilitates calculation of the transport parameters. In addition values of pH and other parameters of tubular fluid can, within certain limits [249, 312], be fixed at desired levels. At the same time variation of the peritubular solution is possible through simultaneous capillary perfusion. Poisons like cyanide, ouabain and others can therefore be administered without risk to the whole animal [293, 295, 306]. This method however is limited by the fact that complete peritubular perfusion cannot be controlled visually for non-superficial portions of the tubule. Complete peritubular perfusion of the tubule can therefore not be guaranteed.

If complete peritubular perfusion is necessary one is restricted at this time to the method of split drops or stationary microperfusion [143, 144, 158, 160, 353]. This latter technique utilizes only short stationary columns of fluid in the superficial segments of the tubule. Adequacy of simultaneous peritubular perfusion can thus be visually controlled. A disadvantage of this technique on the other hand lies for instance in the fact that the influence of rate of flow on transport events cannot be determined. Reabsorption from a stationary drop cannot be compared therefore to transport out of a rapidly flowing fluid (unstirred layers, laminar flow, etc.). An additional source of error arises from the fact that the samples from such a single drop are extremely small; pooling of samples is therefore usually required. ULLRICH and collaborators [342] were able in this manner to prove the dependence of amino acid reabsorption on sodium concentration. LINGARD and collaborators [201, 204] utilized this technique to demonstrate that the ability of the proximal tubule to reabsorb amino acids varies with its distance from the glomerulus.

A different microtechnique was used by BERGERON and MOREL [22]. They injected labeled amino acids into different sections of a superficial nephron and collected urine for determination of radioactivity. This method was later extended also to peritubular capillaries [24, 25]. Several interesting questions have been approached in this manner. Because the rate of injection and flow at the site of injection are usually not known the final conclusion however can never be completely quantitative.

The micromethod which probably interferes least with physiological conditions is the simple free-flow micropuncture. Although it does not permit variation of parameters as in the case of perfusion techniques, it can lead to concentration profiles along the nephron under natural free-flow conditions as shown in the semiquantitative experiments of WEISE et al. [354, 355].

This section provides only a short summary of micromethods used so far *in vivo* for the investigation of amino acid reabsorption. Further technical details may be found in the following reviews: General [82, 340, 362]; continuous microperfusion [15, 86, 309, 319]; split droplet [131, 143, 144, 158, 160, 201, 351]; free-flow puncture [63, 270, 271].

These micromethods have been applied to the study of amino acid reabsorption only in the rat. It is not possible without risk therefore to extrapolate the results to man. LINGARD et al. [203] however have recently published detailed clearance measurements for amino acids in the rat. These show great similarities to values observed in man. Essential differences were seen only in clearances of glycine and histidine. In man these amount respectively to 3.5 and 6% of simultaneous inulin clearance, whereas in the rat they lie below the usual maximum for amino acids of 2%.

In summary it is clear that the choice of a suitable species and an appropriate technique, as discussed in this section, will remain critical for the analysis of various aspects of renal function.

IV. *In vitro* Results

It is likely, as discussed in Section III that solute movement in renal cortical slices primarily reflects events at peritubular cell membranes. Peritubular transport thus studied in vitro generally closely resembles that seen *in vivo*. Thus, similar systems can readily be demonstrated under both conditions [117]. For instance, a separate and specific mechanism in both cases mediates transport of acids of the ALO-group [1]. Again, neither glutamic nor aspartic acid interacts with transport systems for neutral or basic compounds.

Application of classical *in vitro* procedures for the study of amino acid transport was described in detail in the comprehensive review of SEGAL and THIER [285], to which reference may be made for earlier work. Results obtained with such techniques continue to be reported. Although the functional significance of these results must be questioned, the technique can provide much useful information on properties of amino acid transport systems in general. Thus, ROSENHAGEN and SEGAL [263] used slices of human and rat kidney cortex to study the stereospecificity of amino acid uptake. Some interaction occurs between D- and L-lysine but no exchange diffusion between the two isomers could be observed. That efflux from slices is likely to be carrier-mediated may be deduced from the exchange diffusion experiments of SCHWARTZMAN et al. [273]. ROSENHAGEN and SEGAL do not exclude participation of luminal membranes in the processes studied but are careful not to draw close analogies between reabsorption of amino acids *in vivo* and their accumulation in slices.

REYNOLDS et al. [252] described an adaptive control mechanism which can regulate accumulation of certain amino acids in renal cortical slices of newborn rats. The effect becomes apparent upon preincubation of slices at 37° C, and is abolished by inhibitors of protein synthesis. Similar findings have been made also with other embryonic tissues. In other words the observed regulatory effects do not refer to a specific renal mechanism: in agreement with the discussion in Sections III and V the peritubular transport mechanisms studied in slices partake of the properties of amino acid transport in tissues in general.

An interesting comparison of *in vitro* transport of L-methionine by intestine and kidney of rats was reported by BARTSOCAS et al. [17]. Interest in this comparison arises especially from the observation of variable methionine excretion in Hartnup's disease [291], and from the recognition of an isolated methionine malabsorption syndrome [180]. In both tissues the compound was accumulated against a concentration gradient. No interaction between methionine and representative imino, dibasic and branch chain amino acids was noted in kidney. The inhibition of glycine accumulation by methionine reported in this study recalls the similar findings made *in vivo* (Section VI. A.1). Interaction between methionine and glycine contrasts with the inactivity of proline, thought to share a common transport mechanism with glycine [221]. It could not be decided whether methionine is transported by an independent mechanism, or by a transport system shared with other neutral amino acids but for which it exhibits

[1] L-*A*rginine, L-*L*ysine, L-*O*rnithine.

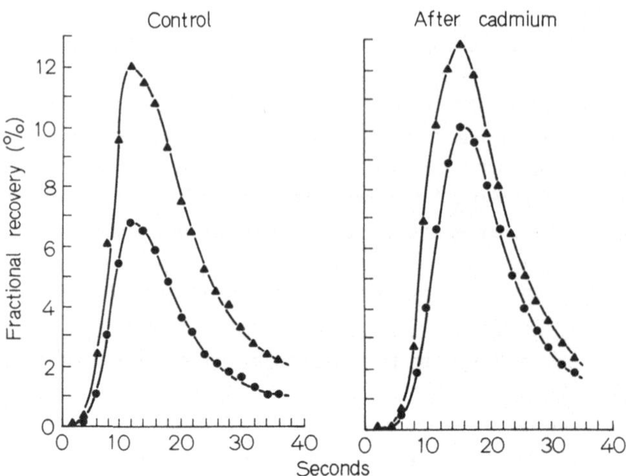

Fig. 1. Peritubular transport of glutamate in rabbits. Reprinted with permission from [117], venous recovery curves are shown for inulin (▲) and glutamate (●); Cd was administered intraarterially immediately after the control period and 20 min before the second study

a much greater affinity. It would clearly be of great interest to study the luminal transport of methionine *in vivo* (see [312a, 312b]).

More recently techniques have been reported which permit the analysis of interaction of various solutes with fragments of luminal or peritubular cell membranes. Primarily so far attention has focussed on sugars [189], but similar approaches have also been applied to the study of amino acid transport [343]. Such experiments may be helpful in localizing transport events on one or the other side of cells.

V. Peritubular Transport of Amino Acids *in vivo*

Peritubular cell membranes in renal cortex can transport a variety of solutes from interstitium into cells without subsequent secretion as shown by SILVERMAN et al. [315] on the basis of results obtained with a double indicator dilution technique. In this method, a bolus containing an extracellular marker like inulin together with a labelled test substance is injected into the renal artery and serial blood samples are obtained from the renal vein. If fractional recovery of inulin in any early fraction exceeds that of the test substance, it follows that the volume of distribution of this substance exceeds that of non-filtered inulin, i.e. the solute must have moved beyond the postglomerular inulin space. As tubular epithelium constitutes the main cell type in the renal cortex it is reasonable to interpret such a finding in terms of interaction with peritubular cell membranes. We will further assume that this interaction leads to cellular uptake and thus reflects peritubular transport.

A typical result obtained with this technique is shown in Fig. 1 [117]. Here the difference between the inulin recovery curve and that for glutamate provides a measure of peritubular uptake of the amino acid. The experiment was carried out before and after administration of Cd (2.5 nM) under suitable conditions [146]. Reduction of the area between the recovery curves reflects the inhibitory effect of the metal on peritubular glutamate transport.

In the rat also, reaction of selected amino acids with peritubular cell membranes can readily be demonstrated [119]. However, no exhaustive study has yet been reported of peritubular transport mechanism *in vivo*. Results so far indicate that in the rabbit possibly 4 separate mechanisms may catalyze movement of amino acids out of the peritubular interstitium [117]. They may be distinguished from one another both on the basis of competition studies, and by their differing sensitivities to compounds of heavy metals. The systems mediate uptake respectively of (a) anionic amino acids, (b) alanine, serine, cysteine, glycine, (c) cationic compounds, and (d) possibly phenylalanine. The possibility cannot be excluded that one and the same system is involved in transport of phenylalanine and other neutral compounds, but than this system possesses a very high affinity for phenylalanine. Only transport of the dicarboxylic, anionic compounds (aspartate, glutamate) is sensitive to inhibition by heavy metals. In addition to the compounds listed, methionine, tryptophan and α-aminoisobutyric acid (AIB) also react with the peritubular membrane but their interaction with other amino acids has not been tested. To the extent at least that system (b) does not react with D-alanine, peritubular transport shows the usual stereospecificity for L-amino acids.

Further work could well lead to the conclusion that additional structural requirements determine the specificity of various peritubular transport systems. Nevertheless significant similarities between peritubular and luminal transport are clearly apparent. In both cases, for instance, separate systems mediate movement of basic and of acidic amino acids. Differences between luminal and peritubular mechanisms have however also been described. This was already observed by SILVERMAN et al. [315] in relation to the specificity of sugar transport at the two sites. For amino acids the metal sensitivity of peritubular systems differs from that of luminal processes [115]. Thus, peritubular transport of aspartate both *in vivo* and in slices could be inhibited by mercurials under conditions where aspartate reabsorption remained normal [115]. An additional difference between transport processes on the two sides of tubular cells has already been referred to in Section III. While luminal transport is primarily involved in reabsorption peritubular transport events may possess a nutritive role. Need for nutritive peritubular uptake would of course be minimal in early sections of the proximal tubule where reabsorption of luminal contents could provide all required substrates. In later sections of the tubule, by contrast, the luminal amino acid level approaches zero and peritubular amino acids constitute the only source of these important metabolites. A similar dissociation of reabsorption at the luminal cell border from nutritive uptake at the peritubular side was suggested for α-ketoglutarate [62] and lactate [93].

The possibility may also be raised that peritubular mechanisms can facilitate efflux of reabsorbed amino acids from cell into interstitium. Whether these mechanisms are capable of active transport has not been unequivocally demonstrated *in vivo*, although inhibition by cyanide and dinitrophenol was reported under conditions where changes in blood flow distribution did not suffice to explain depression of peritubular amino acid uptake [116]. In slices of course the presumably peritubular uptake of amino acids takes place against significant concentration gradients. It is likely that detailed study of these peritubular processes will continue to have to rely to a large extent on the slice technique, as was the case in the past for the analysis of PAH transport [78], and that of sugars [192] and amino acids [255a].

The demonstration of peritubular transport mechanisms for amino acids leads to the conclusion that these compounds may at times be transported in opposing directions across peritubular and luminal membranes. Although the functional significance for such bidirectional transport remains open to speculation, one may recall similar bidirectional movement of other solutes such as sugars [315], uric acid [157] and potassium [113]. In each case, the net transcellular movement of the solutes will depend on characteristics of the two opposite membranes.

VI. Amino Acid Reabsorption

This section discusses reabsorption of amino acids under the separate headings outlined in Section II.

A. Neutral Amino Acids

1. Glycine

At normal plasma glycine concentrations of 0.15 to 0.44 mM [8, 79, 95, 107, 123, 288, 289, 323] man and dog reabsorb 95 to 98% of the filtered load of glycine [138, 323]. In man clearance values for glycine from 0.8 to 6.6 ml/min/1.73 m^2 body surface have been reported. Comparable investigations have also been performed in other species [106, 169, 239]. With increasing plasma levels of glycine the rates of both reabsorption and excretion will rise [138, 243, 323]. Although stop-flow studies clearly indicated that the proximal tubule is a site of major glycine reabsorption [43, 136, 170, 267], micropuncture experiments subsequently showed that several times more glycine remains in tubular fluid collected from the terminal portion of the proximal convoluted tubule than is excreted in the urine [354]. As essentially no reabsorption of glycine takes place in either the distal convoluted tubule or the collecting duct [22], the conclusion becomes necessary that the pars recta and/or other sections of Henle's loop are capable of reabsorbing glycine.

CHRISTOPHEL and DEETJEN [58] were the first to use the microperfusion technique to investigate tubular amino acid reabsorption. They found an approximately exponential decrease of glycine concentration along the length of the proximal convoluted tubule, almost independent of the initial glycine concentration [58, 59]. This reabsorption could be inhibited by addition of L-methionine; L-proline by contrast did not act as inhibitor. Intratubular infusion of 10^{-4}M [2] 2,4-dinithrophenol also remained without effect on glycine reabsorption [59, see also 306]. Unfortunately some of these studies were carried out at a very low pH-value. It became desirable therefore to repeat some of the critical experiments of CHRISTOPHEL [59] at a pH-value of 7.4 [293, 294, 295].

Buffering of the perfusion solution was achieved by addition of 21 mM phosphate. Because of the possibility that this buffer might influence reabsorption of amino acids [12, 99, 213, 323] control experiments were carried out with another buffer system. The biologically inactive compound TES (Tris-(hydroxymethyl)-methyl-2-aminoethane sulfonic acid) [150] was chosen for this purpose. As can be seen from Fig. 2 control values with TES buffer could not be distinguished from those obtained with phosphate buffer. There was thus no reason to expect an influence of phosphate on glycine reabsorption in the proximal tubule under present conditions. It must be emphasized, however, that even though the perfusion fluid may be buffered at pH values higher or lower than about 6.6, this value will not be maintained throughout the entire tubule [249, 312].

A particularly controversial question relating to tubular glycine transport refers to the existence of the so called transport maximum (Tm) [22, 26, 138, 147, 148, 208, 243, 244]. To investigate this problem proximal tubuli were perfused with solutions containing different concentrations of glycine. The results are illustrated in Fig. 2. Over a range of initial glycine levels of 0.2 to 3 mM the concentration along the proximal convoluted tubule decreases at the same fractional rate. In this concentration range unidirectional glycine outflux thus appears to be proportional to the initial glycine concentration. One could assume either that this represents simple diffusion according to Fick's law. Alternatively, if a transport system is involved in this process the maximum velocity (V_{max}) would have to be so high that no saturation level had yet been approached at a concentration of 3 mM.

To help distinguish between these two possibilities the intratubular glycine concentration was increased to 20 mM, an admittedly unphysiologically high level. Results of these studies are also shown in Fig. 2. Starting from this relatively high initial value the percentage rate of decrease of glycine concentration along the tubule is clearly much depressed, a finding incompatible with simple diffusion. No further decrease in the slope of the concentration profile resulted from a further increase of the glycine concentration to 40 mM.

These findings suggest the presence of two components of glycine reabsorption. Only one of these becomes saturated below 40 mM. To determine further whether the second process, unsaturated at a concentration as high as 40 mM, represents energy-linked transport, effects of metabolic inhibitors were next studied. Addition of HCN or 2,4-dinitrophenol on the luminal side caused no depression of glycine reabsorption ($[Gly]_0 = 3$ mM). This finding was surprising: at least

[2] In this paper M is used as abbreviation for mol/l (mM = mmol/l).

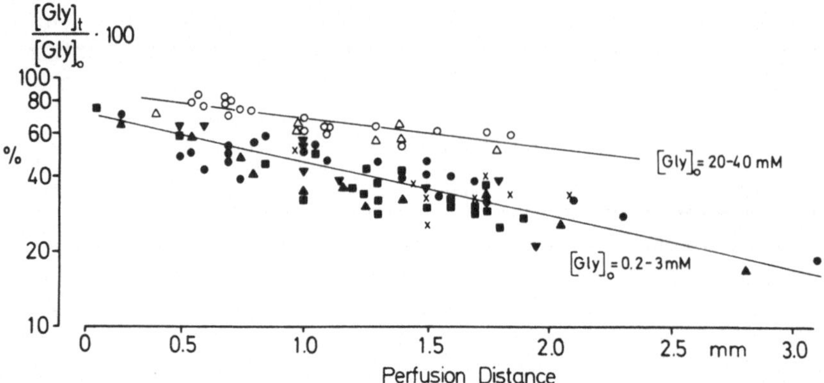

Fig. 2. Relationship between relative ^{14}C-glycine concentration and length of perfused segment (mm), at different initial glycine concentrations $[Gly]_0$, in mM: 0.2 (●), 1.0 (▼), 3.0 (▲), 20.0 (○), 40.0 (△). ■ = TES replaced PO_4-buffer at $[Gly]_0 = 0.2$–3.0. × = addition of 2,4-DNP (0.1 mM) at $[Gly]_0$: 0.2–3.0. Rate of perfusion (Q) 16nl/min (from [295])

the easily saturable first component of glycine reabsorption was expected to show sensitivity to such inhibitors. Accordingly, reabsorption of 2,4-dinitrophenol and cyanide themselves was determined [306]. Diffusion of these compounds out of the tubular lumen proceeded so fast that after a short distance along the perfused tubule the inhibitor concentration had fallen below presumably effective levels.

Attempts were therefore made to study effects of these inhibitors when presented to the tubule on the blood side by peritubular capillary perfusion. The rate of decrease of glycine concentration along the proximal tubule was then clearly slowed below control values following addition of the inhibitors (see Fig. 3). Complete inhibition of glycine reabsorption could however not be obtained.

Presumably reabsorption continuing after administration of HCN corresponds to that portion which showed no sign of saturation at 40 mM glycine. We may assume therefore that this portion of glycine reabsorption can be ascribed to simple physical diffusion. To find further support for this view in some preliminary experiments an attempt was made to abolish transtubular concentration gradients of ^{3}H-glycine by preloading animals with that compound. Under these conditions a much reduced net diffusion was observed.

It seems safe, therefore, to assume that the portion of total glycine reabsorption consists of simple diffusion and that tubular epithelium therefore must exhibit significant passive permeability to the solute. It further follows that Fig. 2 does not describe net glycine transport, but only the unidirectional removal of labeled glycine from the tubule. Estimation of net reabsorption *in vivo* under free flow conditions must take into account the fact that glycine may also diffuse from peritubular interstitium into the tubular lumen. The quantitative significance

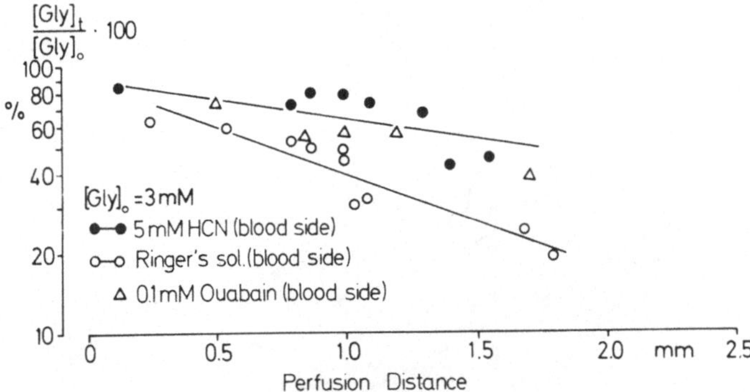

Fig. 3. Effect of inhibitors on glycine reabsorption. Experimental details as for Fig. 2, with addition of peritubular perfusion as shown (from [295])

of such a leak in the direction of secretion is diminished by the large volume reabsorption in the proximal tubule which tends to increase the intratubular concentration of glycine. As a rough estimate the net passive diffusion out of the tubule can account for reabsorption of perhaps 50% of a filtered glycine load.

These results and their interpretation clearly disagree with the concept of a transport maximum described for glycine by PITTS [243, 244, 245]. It is difficult to decide to what extent analytical problems can explain these discrepancies but other clearance studies also failed to reveal a tubular maximum for glycine reabsorption [148]. The micropuncture studies described here thus confirm the latter reports, as well as more recent clearance studies [138].

On the basis of rates of reabsorption measured at initial glycine concentrations of 20 to 40 mM the permeability of the tubule for glycine approximates 13×10^{-5} cm/sec, a value similar to that found recently for L-histidine (202, see also Section VI. 1.d). On the other hand the tubule is far less permeable to L-arginine [297, 300, 302] and L-tryptophan [48]. For these two amino acids active transport systems can explain the major portion of total reabsorption [48, 297, 300, 302].

Extensive work on glycine uptake by slices also suggested the presence of two separate transport mechanisms. HILLMAN et al. [171] concluded from experiments with isolated tubules that three such systems were involved in glycine transport. On the other hand clinical findings suggest the activity of one system for glycine alone, another for glycine and imino acids (see Section VI. A.2). Although glycine metabolism may make interpretation of such results difficult, it was further suggested that movement of the non-metabolizable α-aminoisobutyric acid might also under certain conditions be mediated by two systems [280]. However, as already pointed out, the relevance of such *in vitro* studies to the analysis of glycine reabsorption remains in question. The problem of how many separate transport systems catalyze tubular reabsorption of glycine and α-aminoisobutyric acid (AIB) thus remains unanswered.

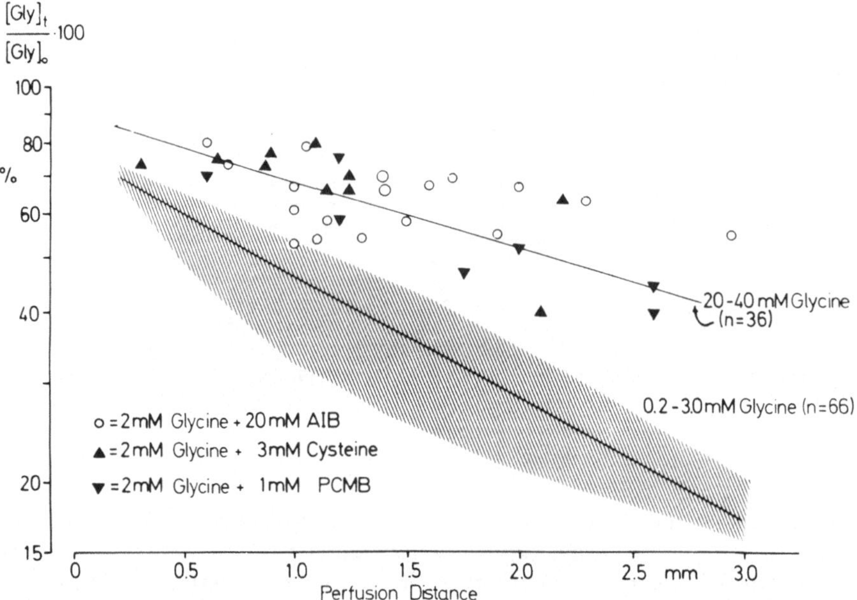

Fig. 4. Effect of AIB, L-cysteine and PCMB on glycine reabsorption ($[Gly]_0=2$ mM). Hatched area and lower line represent control values at $[Gly]_0=2$ mM, upper line those for $[Gly]_0=20$–40 mM (from Fig. 2); $Q=16$nl/min (from [296])

Both competitive and non-competitive inhibitors have been observed to depress glycine reabsorption. As an example of competitive inhibition in Fig. 4 the tubular reabsorption of 2 mM glycine in presence of 20 mM AIB [296] is shown. Compared to control values there is seen a clear diminution of the rate of fractional reabsorption of glycine in the presence of AIB. Values are gathered around the straight line obtained when glycine alone was present in high concentration. This fact suggests that at this level AIB can essentially saturate the active portion of glycine transport. Saturation of glycine disappearance in these studies presumably reflects saturation of a transport mechanism not of glycine metabolism. This follows from the fact that the non-metabolizable AIB in high concentration can depress fractional reabsorption of 2 mM glycine to the same extent as does 20 to 40 mM glycine in the absence of AIB (see Fig. 2).

Results of the reverse experiment [296], in which interference of glycine with reabsorption of AIB was studied, are shown in Fig. 5: The intratubular concentration profile of labeled AIB (0.9 mM) is represented by the open circles and the lower regression line. Upon increasing the AIB concentration to 20 mM a marked decrease in the rate of fractional reabsorption of AIB was observed. The same result is seen when reabsorption of 0.9 mM AIB was measured in presence of 20 mM glycine. These results show that the transport of AIB can be saturated. In addition they demonstrate competitive interaction between AIB

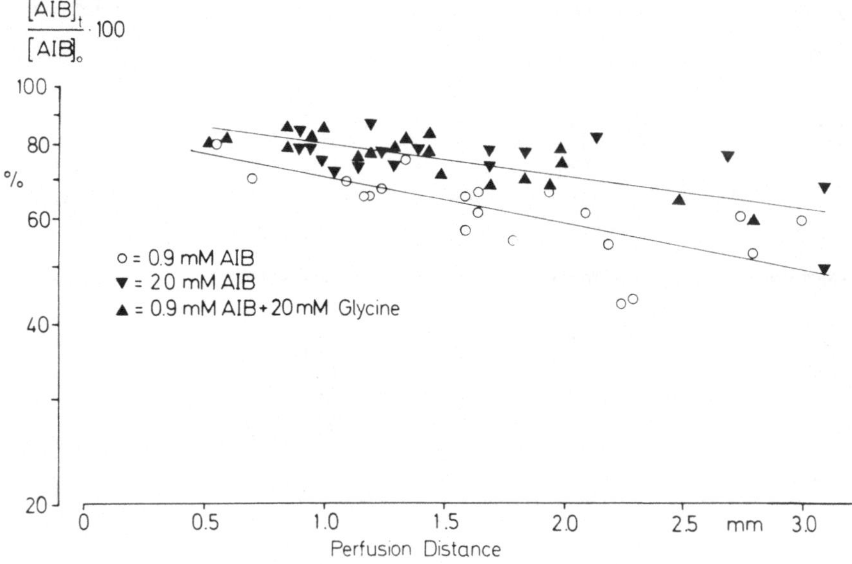

Fig. 5. Saturation and inhibition of AIB reabsorption; $Q = 16$ nl/min (from [296])

and glycine *in vivo*, leading to the conclusion that these two amino acids utilize a common transport system in the tubular cell. These findings are in contrast to earlier results from clearance experiments [351].

The neutral amino acid L-cysteine with its polar side-chain (see Fig. 6) inhibits glycine reabsorption already at a concentration of 3 mM (see Fig. 4). In this case however, the reverse effect, that is inhibition of L-cysteine reabsorption by 20 mM glycine, could not be demonstrated (see Fig. 12, Section VI.A.3). Two possible explanations may account for this findings. Either L-cysteine here inhibits glycine reabsorption non-competitively; alternatively one would have to assume that although a common transport mechanism may exist for these two amino acids, its affinity for glycine is so low that no measurable inhibition is seen even at high glycine concentrations. The inability of glycine to depress cysteine reabsorption may incidentally be contrasted with the very strong inhibition seen in the presence of L-arginine (see Fig. 12).

Another sulfur-containing amino acid, L-methionine (see Fig. 6), also depresses tubular reabsorption of glycine [59, 303, 346]. The nature of this inhibition remains unclear. In any case high concentrations of methionine can be quite toxic [162] and could affect glycine transport completely non-specifically.

In clearance experiments (as also *in vitro*) L-proline, L-hydroxyproline and L-alanine (see Fig. 6) greatly influence transport of glycine [197, 221, 275, 279]. The finding with proline and hydroxyproline recalls the syndrome of iminoglycinuria [14, 103] which can result both from hyperprolinemia as well as from intrarenal causes. On the other hand a hyperprolinemia with prolinuria but no increased excretion of glycine has also been described [222]. In microperfusion

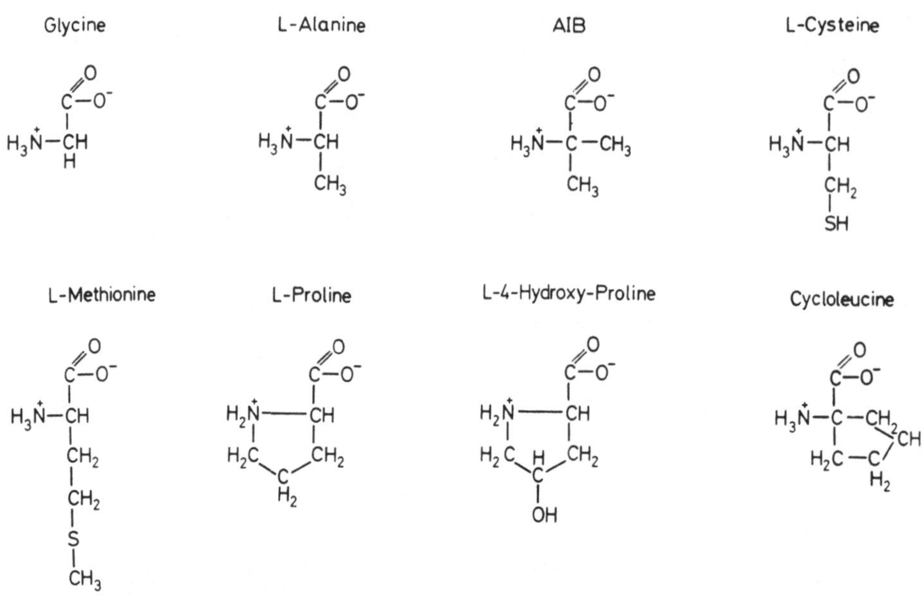

Fig. 6. Structure of neutral amino acids (see also Figs. 16, 18, 25)

studies of the proximal convoluted tubule no influence was seen of L-proline on reabsorption of glycine [59]. These descrepancies remain unexplained.

Some information has been gathered on the source of energy for and molecular mechanism of glycine reabsorption. Thus in tubular and capillary microperfusion experiments cyanide and ouabain inhibited glycine reabsorption [293, 295]. These results suggest that active glycine transport depends on aerobic cell metabolism and in addition that it may be coupled with Na-K-ATPase, for which ouabain is a relatively specific inhibitor. The observation that parachloromercuribenzoate (PCMB) inhibits tubular reabsorption of glycine (see Fig. 4) suggests the possible involvement of free SH-groups. Sulfhydryl reagents such as PCMB also effect electrolyte transport in renal cells [191, 225]. In addition, a number of investigators found that PCMB inhibits Na-K-ATPase of renal tissue *in vitro* (see for instance [190, 232]). Whether PCMB inhibits tubular transport of glycine directly or indirectly remains to be clarified. An effect on sodium and potassium transfer in the tubule cells is possibly involved here. A dependence of glycine reabsorption on simultaneous sodium transport such as was observed *in vitro* [122, 333], has now also been demonstrated in micropuncture experiments by ULLRICH and collaborators [342]. In addition, changes were observed in the transcellular electric potential during amino acid transport [131], a finding which further points to a relationship between amino acid and electrolyte transport.

The increased excretion of glycine together with glucose in the de Toni-Debrè-Fanconi syndrome [31, 32, 84, 92, 108, 109, 110, 200], during lead intoxication [360, 363], in Wilson's disease [33], in a case of lysol intoxication [320], and especially in patients with hereditary glucoglycinuria [184] suggested common elements in the renal transport processes for glucose and glycine. As further

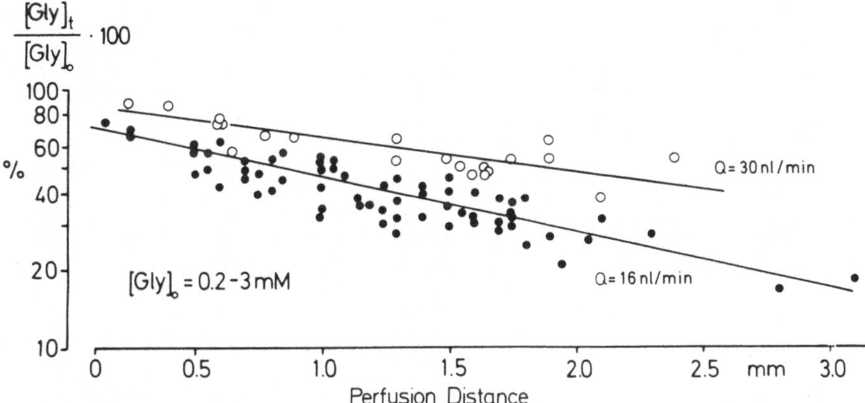

Fig. 7. Influence of flow rate Q on glycine reabsorption; $[Gly]_0 = 0.2–3.0$ mM (from [295])

examples, infusion of glucose decreased the transtubular glycine transport and inversely glycine also depressed glucose reabsorption [349]. However, infusion of sufficient phlorrizin and phloretin to cause glucosuria caused no depression of glycine reabsorption [350]. Reabsorption of glycine and glucose must therefore differ in at least one essential step. The mutual inhibition observed could result also from competition for a common source of metabolic energy [281, 331]. Perhaps finally both transport processes are coupled in some fashion to transport of sodium as mentioned above.

Another interesting effect on tubular transport of glycine is exerted by maleic acid. Injection of maleate leads to a roughly eight-fold rise in excretion of amino acids [5, 6, 7, 166, 233]. In addition glucosuria and phosphaturia are produced so that the functional lesion closely resembles that seen in the Fanconi syndrome. In microinjection experiments no competition between maleic acid and glycine or AIB could be observed [23, 25]. Because maleic acid also inhibits reabsorption of glucose and phosphate it may be inferred that the poison affects a common energy supply for the various transport mechanisms. It is possible also that alterations in renal handling of protons and HCO_3^--ions in presence of maleate [27] could lead to changes in luminal and/or intracellular pH of sufficient magnitude to affect amino acid transport.

Little is known about the influence of flow dynamics along the nephron on reabsorption of glycine. During osmotic diuresis excretion of glycine may increase [147]. Similarly in microperfusion experiments an increased flow rate through the proximal tubule decreases glycine reabsorption [59, 293, 295]. At a low rate of flow (16 nl/min) and at low glycine concentrations an apparent permeability of 22.4×10^{-5} cm/sec was calculated (see Fig. 7).

In contrast, at high glycine concentrations a value of 12.2×10^{-5} cm/sec was obtained as shown in Fig. 8. At both concentrations increasing the flow rate in the tubule to 30 nl/min significantly reduces the rate of fractional reabsorption. However computation of apparent permeabilities yields essentially the same values for the apparent permeability at low and high perfusion rates (see Table 1). Therefore the reduced fractional reabsorption at high flow rate presumably results

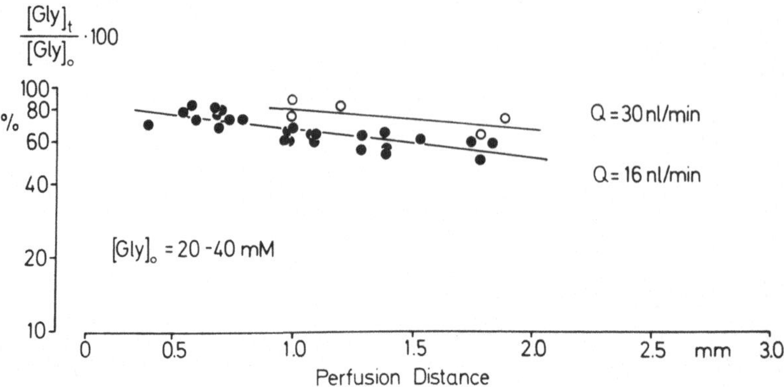

Fig. 8. Influence of flow rate on glycine reabsorption; $[Gly]_0 = 20$–40 mM (from [295])

from the shortened contact time. In addition it should be emphasized that these values obviously can only approach the true permeability coefficient at the higher glycine concentrations at which the active transport component is already saturated.

Table 1. Apparent permeability constants for glycine (P_{Gly}) at different perfusion rates (Q) and initial concentrations ($[Gly]_0$). (r = tubular radius, from Ref. [85])

$[Gly]_0$ (mM)	Q (nl/min)	r (10^{-4} cm)	P_{Gly} (10^{-5} cm/sec)		
0.2–3.0	16	9.4	22.4	$p > 0.05$ (22.4 vs 23.2)	$p < 0.01$ (22.4/23.2 vs 12.2/14.4)
0.2–3.0	30	11.1	23.2		
20–40	16	9.4	12.2	$p > 0.05$ (12.2 vs 14.4)	
20–40	30	11.1	14.4		

Changes in absolute reabsorption because of shortened contact time can also explain the above mentioned fact that the clearance of glycine is increased during osmotic diuresis [147]. Under these conditions a further effect also comes into play: the reduced volume reabsorption during osmotic diuresis leads to smaller increases in glycine concentration along the proximal tubule, and therefore to a lower driving force for passive outward diffusion.

2. Proline and Hydroxyproline

The normal plasma concentration of proline in man ranges from 0.12 to 0.48 mM [8, 14, 79, 95, 96, 107, 275, 323]; hydroxyproline varies around 0.01 mM [14, 277]. The filtered load of these two imino acids is essentially completely reabsorbed so that the normal clearance of L-proline amounts to only 0 to 0.5 ml/min while no hydroxyproline is excreted in urine at all.

In contrast, in cases of hyperprolinemia [101, 102, 104, 153, 193, 220, 222, 241, 269, 274, 275, 286, 316, 366] significant excretion of both imino acids as well as increased excretion of glycine is observed. The prerenal defect underlying this disease leads thus to a so-called overflow aminoaciduria, during which the increased amounts of luminal proline inhibit reabsorption also of hydroxyproline and glycine.

Infusion of L-proline or L-hydroxyproline into normal man also causes an iminoaciduria with increased glycine excretion [275, 277]. The threshold level for L-proline in plasma above which prolinuria is observed approximates 1 mM. At this concentration the system catalyzing reabsorption of L-proline apparently becomes saturated [275].

In these and similar experiments excretion of all other amino acids remained practically unchanged [275, 277, 347]. This fact logically led to the conclusion that a common specific transport mechanism must exist in proximal tubules for the imino acids and glycine [22, 274]. Further support for this interpretation is provided by the specific increase in excretion of imino acids and glycine in cases of so-called renal iminoglycinuria [14, 124, 151, 182, 247, 262, 278, 328, 329, 358]. These patients exhibit normal plasma concentrations of all three compounds. In contrast to iminoglycinuria accompanied by hyperprolinemia we are therefore here dealing with a defect tubular reabsorption.

In Hartnup's disease, another aminoaciduria also of renal origin [16, 103, 214, 215, 217, 276, 356], some of the other neutral amino acids are excreted in pathologically increased amounts; in contrast reabsorption of L-proline, L-hydroxyproline and glycine remains within normal limits. Such facts also are consistent with the existence of a renal transport system specific for these 3 amino acids.

On the other hand a pathologically [50, 235, 272] or experimentally [275, 347] increased concentration of glycine in plasma does not lead to prolinuria or hydroxyprolinuria. Although these findings could be explained on the basis of a very low affinity of glycine for the transport system, occurrence of an isolated renal glycinuria [344] suggests that glycine can also be reabsorbed by a specific tubular mechanism which does not react with L-proline and L-hydroxyproline. It is worth recalling further that an increased filtered load of glycine will reduce reabsorption of several neutral amino acids [275, 347] other than proline. The assumption of a single reabsorption system for glycine and the imino acids is thus probably not justified.

In vitro investigation with isolated tubules [171, 172, 173] yielded some additional information. Evidence was found both for a system possessing high substrate affinity as well as one with much smaller affinity. In addition, existence of a third transport system specific for L-proline was suggested; this system appeared to react also with alanine.

Microinjection experiments *in vivo* [22] showed that L-proline and glycine are reabsorbed in the proximal tubule and further confirmed the mutual interaction between the reabsorption of L-proline and L-hydroxyproline. In agreement with earlier results L-proline was found to possess the higher affinity for this transport system. Saturation of proline reabsorption could not be observed under these conditions [51]. In the microperfused proximal convoluted tubule [59] 21 mM proline did not depress reabsorption of 7 mM glycine.

Other micropuncture experiments [355], in which the amino acids were determined chemically yielded an approximate concentration profile for L-proline along the nephron. At the end of the proximal convolutions there were still present 3% of the filtered load although practically none was found in the final urine. Reabsorption of proline must therefore also involve more distal portions of the nephron [355], and more specifically perhaps the pars recta in Henle's loop, as L-proline injected into the distal tubule is essentially completely recovered in urine [22].

A preparation of brush border fragments from isolated tubules has been described which binds L-proline. Saturation in this case is reached only at the relatively high concentration of 75 mM [174]. Removal of sodium, addition of cyanide, glycine and alanine inhibit this binding whereas 2,4-dinitrophenol and ouabain exert no effect. Whether reaction with such membrane fractions form part of the process of reabsorption is not known. It may further be mentioned that, as in the case of many other amino acids, excretion of L-proline is also increased by cycloleucine. Whether maleate interferes with reabsorption of L-proline and L-hydroxyproline has not been tested.

All these results can be summarized by the statement that L-proline and L-hydroxyproline must use at least one common reabsorption system in the renal tubule. A clear relationship between reabsorption of the imino acids and that of glycine has also been established though many details remain to be clarified. Except in the additional case of L-alanine there is essentially no interaction between transport of imino acids and that of other physiologically occuring amino acids. The assumption of a separate system catalyzing reabsorption of imino acids is therefore probably justified.

3. Cystine and Cysteine

Cystine owes its name to the fact that it was first found over 150 years ago in the form of stones and hexagonal crystals in the urinary bladder [28, 234, 364]. This pathological condition was therefore referred to as cystinuria. Early this century the chemical structure of cystine was established [126] and cystinuria was considered as consequence of an inborn error of metabolism [134, 206]. Unsatisfactory analytical methods for cystine and the other amino acids in plasma and urine contributed greatly to the fact that for many decades no further progress was made in elucidating the pathogenesis of cystinuria. Thus it was difficult for instance to distinguish between cystinuria and the cystine storage disease or cystinosis [1, 200].

In 1912 lysine was found in the urine of cystinuric patients [1]. Thereupon improved analytical methods for amino acids revealed in addition an increased arginine and ornithine excretion [324, 369]. Further, small amounts of a mixed disulfide homocysteine-cysteine were also found (see for instance [128]). Confirmation of the view that the cause of cystinuria lies in a renal defect [37, 38] was later provided by the work of DENT and ROSE [87, 88]. These authors found in cases of cystinuria normal or even reduced plasma concentration of the amino acids involved. Because of the structural similarity of L-arginine, L-lysine, L-ornithine (ALO-group) and L-cystine these authors suggested for the first time a common transport system for these amino acids in the kidney (It must be added that a pathologically increased excretion of L-ornithine in cases of cystinuria had not been demonstrated at that time). According to this view cystinuria represents malfunction of one specific tubular reabsorption mechanism.

These fundamental studies led to a more detailed analysis of the renal handling of the ALO-group and cystine. The main focus of most of this work remained on the problem of cystinuria [97, 121, 163, 164, 170, 194, 248], and also reviews [18, 20, 77, 103, 141, 142, 165, 217, 334, 335, 372].

The plasma concentration of L-cystine ranges from 0.008 to 0.062 mM [8, 79, 107, 275, 323, 325], and reliable values for its clearance range around 3 ml/min/ 1.73 m^2 of body surface [79, 275, 323]. Absolute excretion of course varies as in the case of other amino acids with protein intake and also with age [19]. It must be emphasized that in most cases L-cystine and L-cysteine could not be separated and were measured together. Reliable data referring to cysteine alone have only occasionly been reported [41].

In stop-flow experiments reabsorption of L-cystine, like that of most other amino acids, was observed in proximal tubular fractions [267]. In clearance studies infusion of L-alanine, L-phenylalanine and L-methionine [82, 347] increased excretion of cystine/cysteine. Cycloleucine also inhibited L-cystine transport [178]. These results all point to some relationship between excretion of cystine/cysteine and that of other neutral amino acids. In the case of L-methionine metabolic breakdown to cysteine/cystine may simulate changes in cysteine/cystine reabsorption [133].

Starting from the theory that the ALO-group and L-cystine are carried by a common transport mechanism in the tubule [87, 88] further work focussed mainly on the interaction between those four substances. A special difficulty in this work relates to the poor solubility of cystine [34, 90, 105, 268]. Indeed in cystinuria it is precisely this property which causes stone formation in the urinary tract. It is interesting to mention that penicillamine which forms a more soluble disulfide with cysteine is used therapeutically in the treatment of cystinuria [68, 69, 72].

Infusion of one of the basic amino acids causes a condition very similar to cystinuria with greatly increased urinary excretion of the ALO-group and of L-cystine [166, 266, 267, 346]. L-cystine was not used in the inverse experiment, presumably because of its low solubility; L-cysteine was infused instead and led to some small increase in excretion of ALO-group acids [346]. Existence of a common transport system for the ALO-group and L-cystine was considered proven on that basis [88, 135, 165, 219, 266] in spite of the absence of convincing evidence for an inhibition of tubular transport of the ALO-group by L-cystine.

Following the discovery that in cases of cystinuria intestinal absorption of the ALO-group and of L-cystine are also depressed [9, 216] a number of investigators took advantage of the greater accessibility of intestinal compared to renal tubular epithelium and studied transport of these four amino acids out of the gut [4a, 161, 210, 211, 228, 250, 251, 258, 259, 330, 332]. It seemed attractive to consider the malfunction in kidney and intestine as manifestations in different organs of one and the same functional lesion of membrane transport, especially in the light of the many similarities in transport by the two organ systems [195, 226, 227, 229, 230, 254, 261].

Of interest in this connection is also the cystinuria which occurs in dogs [39, 40, 64, 67, 152, 155, 336, 337, 338]. This could perhaps serve as a model for the human disease although it is not accompanied by malfunctioning intestinal absorption of the ALO-group or L-cystine [117, 179, 339]. The reported occurrence of isolated cystinuria [81] in the Kenya genette has not been confirmed [70]. On the other hand a specific increase in cystine excretion in man with apparently normal reabsorption of the ALO-group has been reported [42]. This finding also cannot be reconciled with the postulate of a single reabsorptive system for these four amino acids.

Many of these investigations however failed to consider the role of L-cysteine. The probable significance of L-cysteine in L-cystine transport [129] and thus also in cystinuria was established by the finding that 70 to 80% of L-cystine taken up by tubule cells is present as L-cysteine [73, 74, 76, 282]. A similar observation was made also in the intestine [261]. It seems very likely that L-cystine is transported at least in part as L-cysteine. Separate transport systems for these two amino acids could have been inferred from differences in rates of maturation of their transport in the new-born rat [284], were it not for the fact that this could reflect simply maturation of enzymes catalyzing interchange between cystine and cysteine. The same enzymes may also play a role in the intestine where cystinuria is accompanied by decreased absorption of L-cystine but not of L-cysteine [258, 261, 314].

Intracellular transformation of L-cystine into L-cysteine thus probably plays a role in cystine transport. Nevertheless it remains difficult to explain in this manner all the findings apparently at variance with the theory of DENT and ROSE [87, 88]. At least two transport systems must be postulated: one for the renal reabsorption of the ALO-group, the other one for cystine/cysteine. The observation that in new-born rats the capacity for renal reabsorption of L-lysine and L-cystine develop at different rates [283] further supports this view. On the other hand we cannot be dealing with two completely independent transport mechanisms. Indeed, as already mentioned, infusion of basic amino acid influences also the excretion of L-cystine.

To explore in greater depth the process of tubular reabsorption of cystine/cysteine, more recent studies utilized microperfusion of single tubules *in vivo* [299, 301, 304, 307]. In these experiments the initial L-cystine concentration of 0.4 mM was found to decrease exponentially along the tubule to approximately 6% of the initial concentration after a tubular length of 3 mm (Fig. 9).

Addition of 2 mM L-lysine, L-cysteine or L-arginine greatly reduced the rate of fractional reabsorption of L-cystine; the strongest inhibition was achieved with L-arginine. In this case at the end of 3 mm the perfusion solution still contained about 80% of the original L-cystine. This corresponds to a diminution of fractional reabsorption over this distance from a control value of 94% to 20% (see Fig. 9). Arginine inhibition does not seem to be very stereospecific, as similar effects are seen with D-arginine, as well as L- or D-penicillamine, an amino acid resembling cysteine. These compounds depress cystine transport to about the same extent as does L-lysine [307]. Inhibition of L-cystine reabsorp-

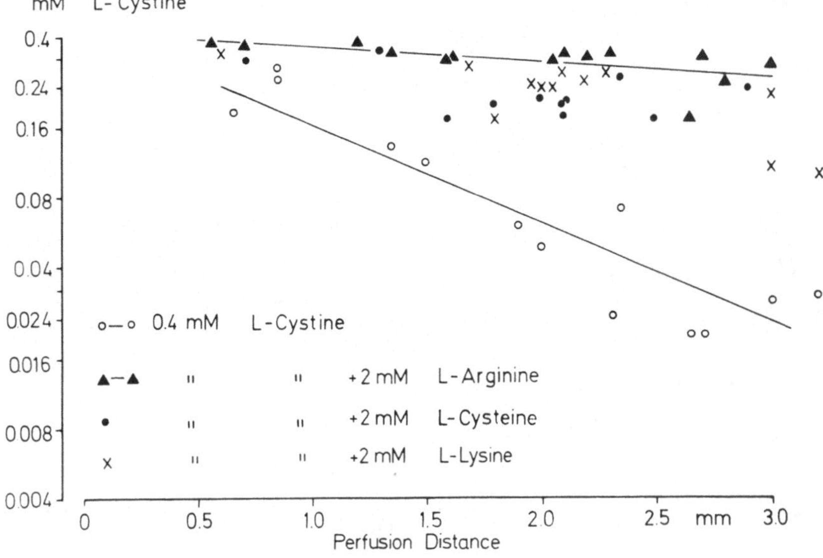

Fig. 9. Reabsorption of L-cystine. Q=20nl/min (from [301])

tion by L-lysine and L-arginine recalls the same effect seen in clearance experiments [255, 266, 267, 346]. The significance of the inhibition exerted by L-cysteine will be further discussed below.

The inhibitory action of L-arginine and L-lysine exhibits some specificity, as can be seen from the results in Fig. 10 (from Ref. [301]): neither decarboxylated arginine (agmatine) nor glycine are able in a concentration of 2 mM to influence the unidirectional flux of labeled L-cystine out of the tubular lumen. Similar results were obtained with L-phenylalanine [313]. Another related compound is 2,6-diaminopimelic acid which possesses the same structure as L-cystine with the exception of a CH_2-group substituted for the S-S-group; such substitution prevents splitting of the molecule into two parts as occurs during reduction of cystine to cysteine. Because this diaminodicarboxylic acid and cystine appear to be transported by a common system in E. coli [21] their possible interaction in the kidney was also investigated. No effect of diaminopimelic acid on tubular reabsorption of cystine could be observed [301]. Together with the inhibition of cystine reabsorption by cysteine illustrated in Fig. 9 this finding provides additional support for the belief that cystine may be largely transported as the split product cysteine.

One would predict on that basis that L-arginine should inhibit reabsorption not only of L-cystine but also of L-cysteine. Fortunately experiments with cysteine *in vivo* possess the advantage that the autoxidizable sulfhydryl compound need not be exposed to the high partial pressure of O_2 commonly employed *in vitro* [73]. At the lower p_{O_2} encountered *in vivo* cysteine dissolved in perfusion solution remained stable for at least three days (Fig. 11). When freshly prepared ^{14}C-

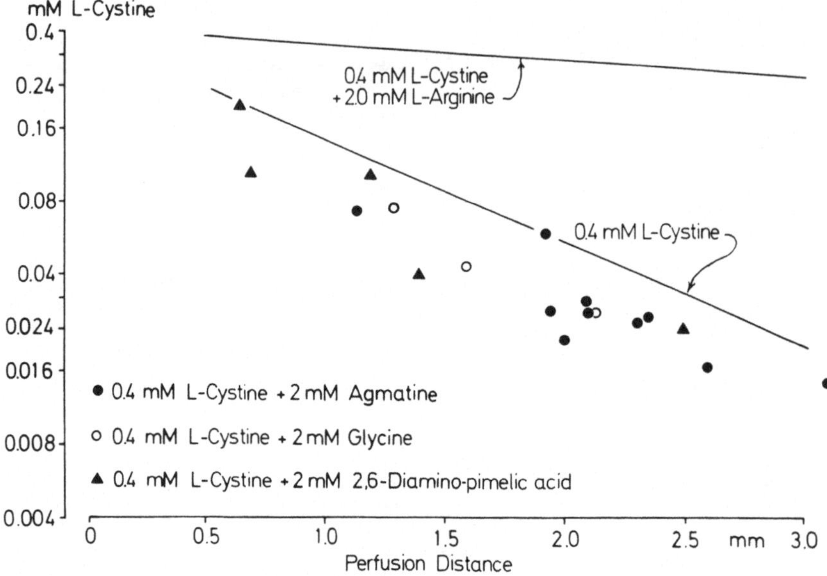

Fig. 10. Reabsorption of L-cystine. The straight lines are derived from Fig. 9 (from [301])

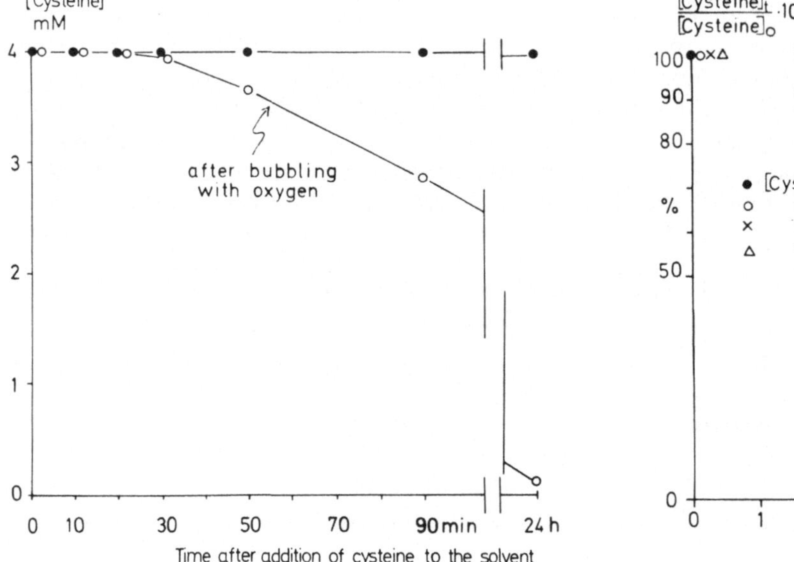

Fig. 11. Stability of L-cysteine in perfusate (from [301])

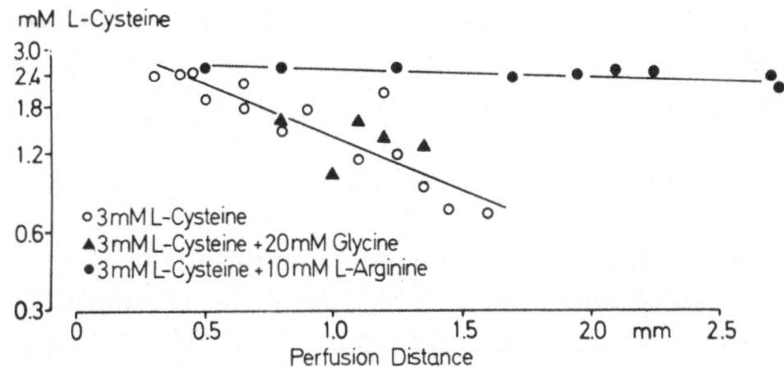

Fig. 12. Reabsorption of L-cysteine. $Q = 16$nl/min (from [301])

cysteine was added at an initial concentration of 3 mM, results were obtained closely resembling those observed with cystine (see Fig. 12) [301]: (note however the slower rate of perfusion than in Fig. 9). L-arginine (10 mM) almost completely blocked transport of L-cysteine (Fig. 12). As control, an even higher concentration of glycine (20 mM) exerted no effect on reabsorption of cysteine. These results correspond exactly to the findings made with L-cystine and provide further support for the view that L-cystine is transported as L-cysteine.

A question critical to pathogenesis of cystinuria now arises: can the micromethods used *in vivo* also demonstrate the reverse effect, namely an influence of L-cystine on reabsorption of amino acids of the ALO-group. Because of the low solubility of L-cystine at physiological pH values this problem can be investigated only with L-cysteine; the process of L-cysteine reabsorption, as shown above, appears identical with that of L-cystine. Accordingly reabsorption of 2 mM L-arginine was measured in presence and absence of 20 mM L-cysteine (Fig. 13). In spite of the high L-cysteine concentration no inhibition of arginine reabsorption could be observed [301].

A similarly negative result was obtained with 2,6-diaminopimelic acid. Even if we assume that cystine/cysteine, or 2,6-diaminopimelic acid exhibit relatively little affinity for the hypothetical transport system they may share with the ALO-group, one would still have expected some effects in the presence of high inhibitor concentrations. An additional finding also speaks against the concept of a com-

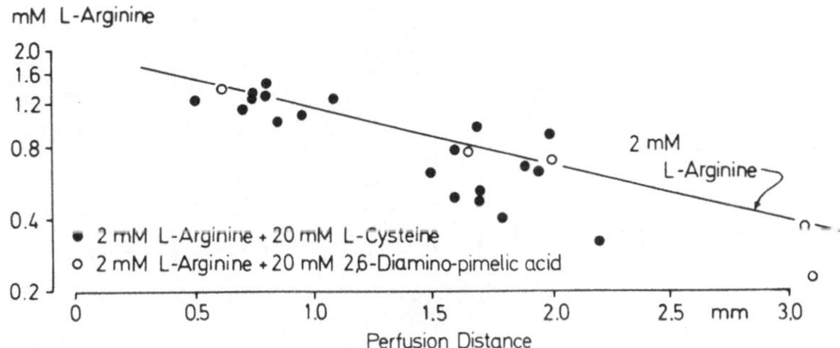

Fig. 13. Tubular reabsorption of L-arginine. $Q = 20$nl/min. The straight line is based on results shown in Fig. 22 (from [301])

mon reabsorption system for L-arginine and L-cystine. As will be shown below, agmatine strongly inhibits reabsorption of L-arginine (see Fig. 26a). However it does not affect reabsorption of L-cystine (Fig. 10). Although different concentrations of the amino acids were used in these two experiments the findings nevertheless strongly argue in favour of separate transport systems for L-arginine and L-cystine [301].

Arguments against the theory of a single common transport system as originally proposed by DENT and ROSE [87, 88] may thus be summarized as follows:

a) cystine/cysteine do not inhibit transport of the ALO-group;

b) there does exist an isolated excretory defect for L-cystine [42];

c) agmatine depresses reabsorption of L-arginine but not of L-cystine [300, 301].

It is necessary to reconcile these findings with the further observations that

d) in classical cystinuria (in homozygotes) both the ALO-group and cystine are excreted in increased amounts;

e) in clearance experiments [255, 266, 267, 346], as well as in micropuncture studies [301], members of the ALO-group inhibit reabsorption of cystine/ cysteine.

The following scheme for the tubular reabsorption of the five amino acids here considered will serve as basis for further discussion [299, 301] (see Fig. 14, left portion): Existence of two transport mechanisms may be assumed of which one (A) mediates transport of ALO-group amino acids but does not react with cystine/cysteine. A second system (B) catalyzes reabsorption of cystine/cysteine but may be inhibited by members of the ALO-group. Whether system B can in addition catalyze transport of the ALO-group across cells remains unknown. Such a possibility does not seem unlikely in light of the existence of two transport systems for L-arginine [220, 224] (see Section VI.C).

The central portion of Fig. 14 represents the situation in "classical" cystinuria. Here system A is thought to be malfunctioning (dashed line) and the tubular concentration of the ALO-group decreases only little along the tubule. As a consequence, system B is inhibited and increased amounts of all the amino acids involved will be excreted. The lesion in cases of isolated cystinuria [42] is believed

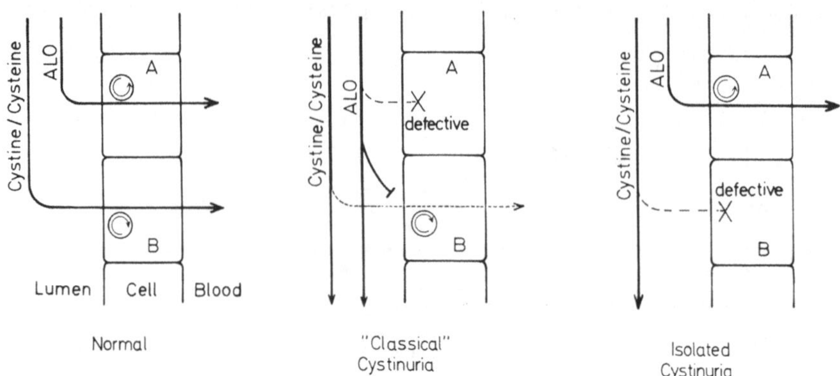

Fig. 14. Model for tubular transport of ALO-group and cysteine/cystine under normal conditions in classical cystinuria and in isolated cystinuria. For details, see text (modif. from [301])

to lie at the level of system B (dashed line, Fig. 14 right portion). This leads to pathological excretion of L-cystine but has no effect on transport of members of the ALO-group.

The suggested mechanism for the renal reabsorption of cystine/cysteine, L-arginine, L-lysine and L-ornithine can account for the five findings (a to e) summarized above. One further fact however has been reported which does not fit into the suggested scheme: certain patients apparently show an isolated defect of tubular reabsorption of the ALO-group in absence of changes in excretion of L-cystine [357, 237]. Detailed analysis however of the data on amino acid excretion in these cases raises some unanswered questions. Thus the excretion of cystine is higher than has been observed normally (for review see [102]); further, 30% of the values in patients exceed the controls presented in the same report [397]. In addition, at normal levels of cystine excretion that of lysine overlaps with values considered normal by other authors [8]. The pattern of amino acid excretion in almost all cases of this so called dibasic aminoaciduria suggests that we are dealing here with heterozygotes of classical cystinuries [15, 223, 260]. This conclusion is strenghtened by the wide quantitive variations in amino acid patterns [20, 75, 103, 165, 186, 187, 217, 240, 334]. It would seem premature at this time to refer to a specific isolated dibasic aminoaciduria.

Another problem so far not mentioned is that of the possible secretion of L-cystine. In clearance experiments with cystinuric patients there have been found occasional clearance values for cystine exceeding that of inulin; in addition, renal extraction of cysteine was observed. These facts suggested the possibility of secretion [71, 75, 127, 223]. Attempts were made to obtain direct evidence for secretion of L-cystine by means of simultaneous microperfusion of the tubular lumen and of peritubular capillaries. For this purpose the peritubular solution

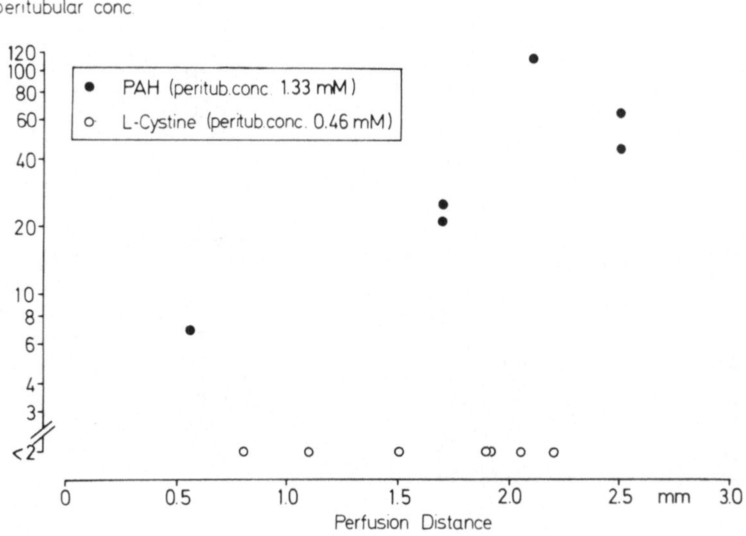

Fig. 15. Absence of L-cystine secretion (from [304]). For details, see text

contained labeled L-cystine while the tubular lumen was perfused with an equilibrium solution containing 20 mM L-arginine. Addition of this amino acid served the purpose of inhibiting reabsorption of any secreted L-cystine. As can be seen from Fig. 15 [304] no measurable L-cystine secretion could be found. In control experiments the well-known secretion of PAH could readily be demonstrated. Although this method may not be able to yield more than semiquantitative results, the results nevertheless fail to support the occurence of L-cystine secretion in the healthy rat. To what extent such a result can be extrapolated to the human cystinuric patient remains an open question. In any case no conclusion is possible at this time on either significance or mechanism or localization of the hypothetical secretion of L-cystine.

4. Other Neutral Amino Acids

Reabsorption of the neutral L-amino acids tryptophan, phenylalanine, tyrosine, alanine, valine, leucine, isoleucine, serine, threonine, asparagine, methionine and glutamine is decreased in cases of Hartnup's disease ([16], see also review and bibliography in [141, 217]). Common steps are apparently involved in the tubular transport of these amino acids but, in spite of many attempts, evidence for a specific reabsorption system for the whole group has not yet been obtained.

Tubular transport of L-tryptophan and L-phenylalanine (for structures see Fig. 16) has been investigated in somewhat greater depth than that of the other compounds. Clearance of endogenous tryptophan in man amounts to 0.7 to 2.0 ml/min [95] at a plasma concentration of 0.06 mM. Values quoted for the

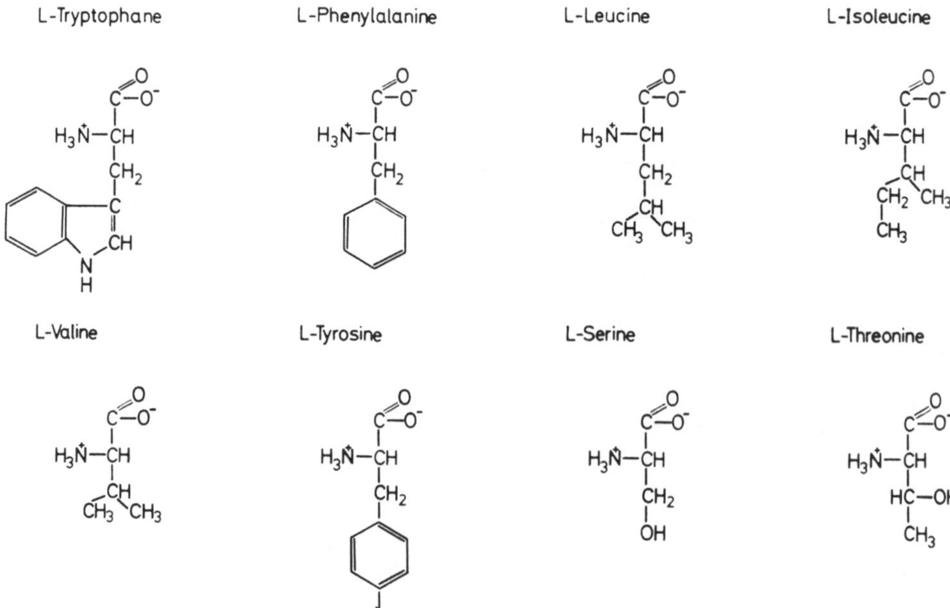

Fig. 16. Structure of neutral amino acids (see also Figs. 6, 18, 25)

clearance of L-phenylalanine range from 0.2 to 1.7 ml/min/1.73 m^2 [8, 79, 95, 107, 275, 288, 323] at plasma concentrations of around 0.02 mM. When the two amino acids were infused in order to increase their renal load, both excretion and reabsorption were raised [96]. No tubular transport maxima could therefore be established for tryptophan and phenylalanine, in agreement with earlier results [29, 30, 265].

In isolated tubule fragments tryptophan is accumulated against its concentration gradient [359] but to what extent this represents movement in the direction of reabsorption is not known. Phenylalanine and 2,4-dinitrophenol inhibit this accumulation. The transport mechanism appears stereospecific, as no intracellular accumulation of D-tryptophan was observed. On the other hand N-acetyl-L-tryptophan is capable of reacting with the system. The same authors also found a secretion of L- and D-tryptophan into the tubular lumen which could be inhibited by probenecid [49, 181].

More recently microperfusion experiments [48] demonstrated the existence of a saturable stereospecific reabsorptive system for L-tryptophan. Here also phenylalanine acted as inhibitor, presumably by competition with tryptophan. The K_m for unidirectional reabsorption of labeled L-tryptophan was reported as 4 mM, and the maximum velocity as 2×10^{-9} mol/cm^2/sec.

Reabsorption of L-phenylalanine was similarly measured [60, 311] and the same saturation characteristics were seen as for tryptophan [48]. In addition, the latter compound strongly depresses phenylalanine reabsorption [60, 311] (see Fig. 17). These findings all support the assumption of a common stereospecific transport system for L-tryptophan and L-phenylalanine. It is less clear whether other amino acids, and especially neutral compounds, can also utilize this system. From clearance experiments it is known [11, 183] that excretion of phenylalanine (and also of tyrosine) is greatly increased following infusion of leucine and methio-

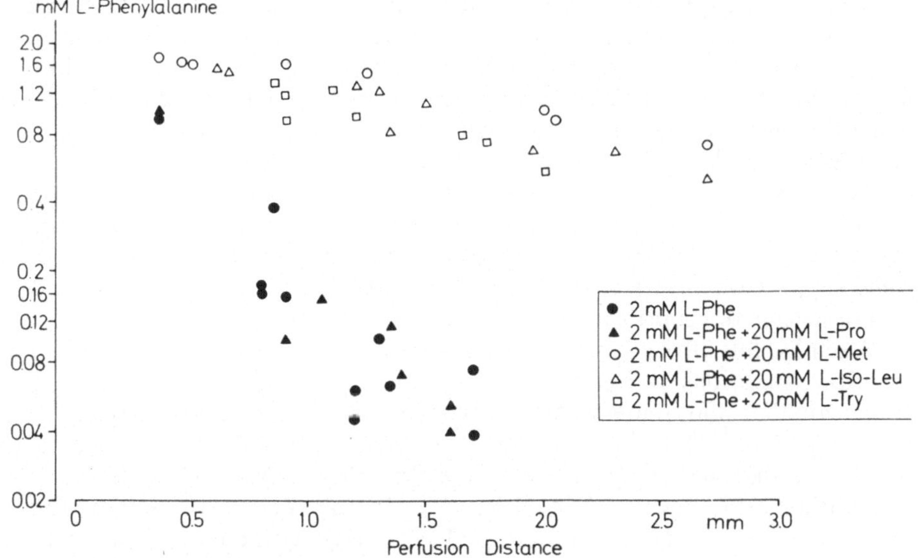

Fig. 17. Reabsorption of L-phenylalanine (from [311])

nine; alanine exerts little effect and no increase was seen following administration of glycine, proline and the anionic amino acids.

These observations could in general be confirmed in microperfusion studies (see Fig. 17, in ref. 311). Here again isoleucine and methionine strongly inhibited reabsorption of phenylalanine, whereas proline remained without effect. Unless proline and phenylalanine possess greatly differing affinities for a hypothetical common carrier, absence of competition between these two compounds argues against existence of such a common system. The nature of the inhibition exerted by isoleucine and methionine will remain unclear until further studies can test the possibility of true competition[3]. The increased excretion of glycine, alanine, threonine, serine, histidine and valine following infusion of phenylalanine also requires explanation. It may be added that phenylalanine exerts no effect on reabsorption of cystine in microperfusion studies [313]. Cycloleucine, another neutral compound discussed further below, and phenylalanine mutually inhibit one another's reabsorption [308, 310, 311], justifying the preliminary assumption that they utilize the same tubular transport system.

The plasma concentrations of *L-leucine* and *L-isoleucine* (see Fig. 16) in man have been reported as 0.08 to 0.18 mM and 0.04 to 0.11 mM respectively [8, 79, 107, 275, 323, 325]. On the basis of more recent chromatographic measurements their clearance values range from 0.2 to 0.9 ml/min/1.73 m^2 [79, 275, 323].

The question of a tubular maximum for leucine reabsorption has not been completely resolved [29, 30, 100]. Of course, the earlier work had to overcome the difficulty that only unsatisfactory methods were available for estimation of amino acids. Similar methodological problems also obscure the significance of clearance experiments in acidosis from the same period [208]. In any case more recent clearance studies failed to establish a tubular transport maximum for leucine [138].

Infusion of L-alanine, L-cysteine, L-methionine or L-phenylalanine (unlike that of glycine and L-proline) inhibits reabsorption of L-leucine and L-isoleucine [174, 347]. L-leucine also strongly depresses reabsorption of L-isoleucine [174, 347]. In microinjection experiments [22] similar findings were made; in addition, the inability to saturate L-leucine reabsorption could be confirmed. Extrapolation of this finding to normal conditions is difficult however. The possible contribution of diffusion to the reabsorption of L-leucine could obscure actual saturation of the transport system, especially in light of the demonstration of a flux of L-leucine from blood to lumen [26]. The fact that the relationship between the load of L-leucine and its rate of reabsorption is not entirely linear [24] offers some support for the view that at very high L-leucine concentrations at least part of leucine transport can be saturated.

A semiquantitative concentration profile for L-leucine and L-isoleucine under free-flow conditions along the nephron has recently been reported [354, 355]. It appears that both these amino acids are reabsorbed relatively quickly in the proximal tubule. Only 20% of the filtered amino acid load remained in the tubule at a distance of 1 mm from the glomerulus; at the end of the proximal convolutions this figure fell to about 3%. In the final urine only a fraction of 1% was recovered. The authors [354, 355] suggest therefore that, as in the

[3] For very recent results concerning this problem see [312a, 312b].

case of several other amino acids, some leucine and isoleucine reabsorption must also occur in more distal portions of the tubule. Presumably this involves the loop of Henle including the pars recta as leucine injected into the distal convoluted tubule is recovered essentially completely in urine [22].

Interesting findings were also made in microinjection studies [25] which helped clarify the action of maleic acid on transport of L-leucine. Maleic acid inhibits net reabsorption of L-leucine in neither a competitive nor a non-competitive manner; instead it apparently increases the permeability to L-leucine in the opposite direction. One can picture a situation in which L-leucine continues to be reabsorbed normally after treatment with maleate while at the same time passive backleakage increases. Accordingly the endproximal equilibrium concentration of L-leucine is raised. Unless however we are dealing here with an anisotropic membrane or even an active secretory mechanism it should be possible to demonstrate the change in passive permeability also in the direction of lumen to blood. This test for the validity of the hypothesis has not yet been performed.

Histidine possesses two amino groups: the α-amino group with a pK-value of 8.97 and a secondary amino group in the imidazole ring with a pK-value of 5.97 (156, see Fig. 18). The compound is therefore related to the basic amino acids; discussion at this stage is mandated however by the fact that histidine reabsorption strongly resembles that of the neutral amino acids.

In the first clearance experiments [98] a reabsorption of 99% of filtered histidine was observed. More recent chromatographic measurements yielded for man a histidine clearance of 7 to 12 ml/min/1.73 m^2 [8, 79, 107, 323, 325] at plasma concentrations of 0.05 to 0.16 mM [107, 275, 323]. Increased concentrations of histidine in plasma obtained either experimentally [138, 371] or in cases of histidinemia [10, 46, 47, 61, 65, 83, 145, 175, 196, 197, 290, 345, 365] led to additional excretion of this amino acid; fractional reabsorption however remains practically constant. No tubular transport maximum could thus be determined in man or dog [30, 138, 327, 368]. In the rat however the possibility of such a tubular maximum cannot be excluded [401]. Differences between histidine excretion in man and rat [203] are also suggested by the reported clearance values: while in man that of L-histidine amounts to around 6% of glomerular filtration rate, this value in the rat is less than 1%.

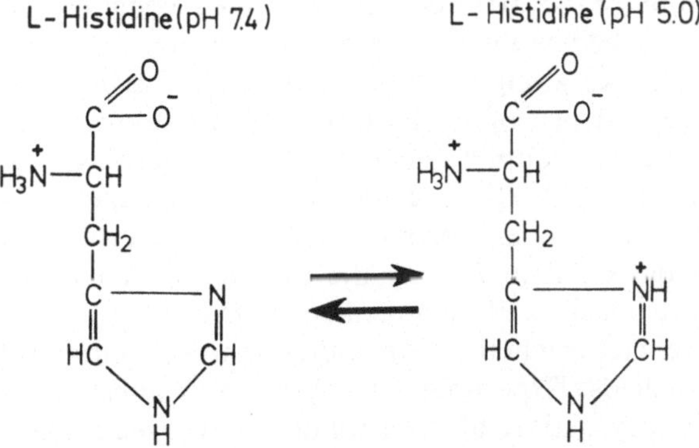

Fig. 18. Structure and ionization of histidine (pK of imidazole group 5.97)

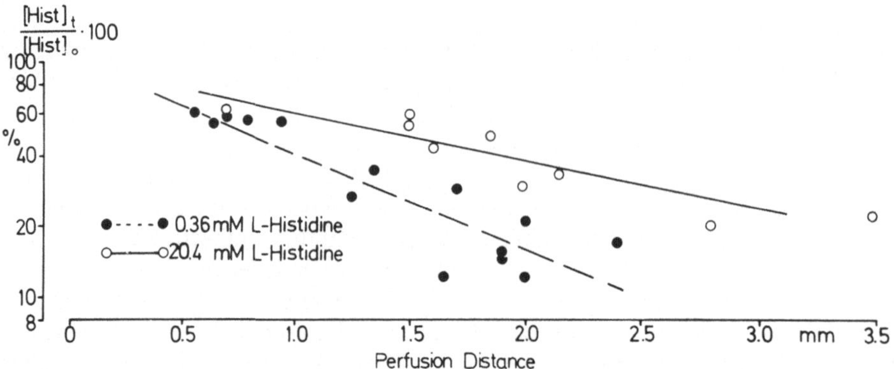

Fig. 19. Reabsorption of L-histidine. ($Q = 20$nl/min) (from [305])

Attempts were also made to determine the rate of tubular reabsorption of histidine in microperfusion experiments [125]. Unfortunately the length of perfused tubule was not reported in these studies, so that the reabsorption rates found possess only limited usefulness. No effect of cyanide on histidine reabsorption could be observed. This was interpreted in terms of a primary importance of diffusion in the process of reabsorption [125]. The rapid disappearance of cyanide from the tubular lumen [306] however may require revision of such an interpretation.

In other similar studies the fractional reabsorption of histidine decreased only from 90 to 70% over a perfusion distance of 2.5 mm when the initial tubular concentration was raised from 0.36 to 20.4 mM (Fig. 19). Excluding this smaller and presumably saturable component the significant fraction of total reabsorption still seen at higher histidine concentrations must consist either of diffusion or of an active transport system for histidine with a very high capacity. The most recent micropuncture results are also compatible with both possibilities [201, 202]. It is interesting that reabsorption of L-histidine in the early proximal tubule differs from that observed in the later proximal portions of the nephron. This finding will stimulate renewed discussion about localization of reabsorption of amino acids in the proximal tubule [141]. As both the maximum velocity (V_{max}) and the permeability (14×10^{-5} cm/sec) remained constant, the origin of this inhomogeneity may be sought in a perhaps as much as eight-fold increase of the affinity constant K_m along the tubule [201, 202]. Further similar experiments demonstrate a dependence of histidine reabsorption on luminal sodium concentrations [341]; actual coupling of transtubular amino acid and sodium transports was postulated [130]. Direct evidence has in the meantime been obtained for a relationship of Na^+-reabsorption with that of certain other amino acids [342]. Such findings confirm *in vivo* the interaction between amino acid and Na^+-transport previously well-documented *in vitro*.

It is not entirely clear whether L-histidine shares a common transport system with other amino acids. Experiments to answer this question led to observation of a small inhibitory action of L-histidine on the reabsorption of basic and some neutral amino acids [174, 176, 346]. Whether histidine can also influence tubular transport of L-glutamic acid remains uncertain [138, 174, 347]. Reabsorption in turn is sensitive not only to imino acids but to almost all neutral and basic amino acids (including L-methyldopa) [174, 347, 370, 371]. Other studies could demonstrate no mutual interaction between the transport of L-arginine and L-histidine [256]. Quantitive interpretation of some of these experiments is rendered difficult because at times [174, 346] only the total L-amino acid nitrogen was measured, rather than the concentration of individual amino acids.

Stop-flow experiments [136] showed in addition that L-histidine and the other basic amino acids are not reabsorbed in the same portion of the proximal tubule. This could reflect existence of two distinct and spatially separated transport mechanisms for L-histidine on the one hand and the ALO-group on the other. Alternatively we might postulate that arginine, lysine and ornithine inhibit transport of histidine as a result of their higher affinity for the transport system. They would therefore be reabsorbed faster than histidine while the latter would only be removed towards the end of the proximal tubule.

A third possibility implicates the decreasing pH along the proximal tubule which might favor histidine reabsorption in later portions of the tubule. The previously quoted pK-value of the imidazole amino group of histidine at 5.97 lies below the physiological pH range of urine in proximal tubules. Nevertheless a drop of pH from 7.4 in glomerular filtrate to 6.8 towards the end of the proximal convolutions entails an approximately four-fold increase in the degree of ionization of this amino group. The ionized configuration meets most of the structural criteria found essential for inhibition of arginine transport (see Section VI.C).

Such facts may explain interaction of histidine with transport of both basic and neutral amino acids. As further test of the explanations offered, the influence of 20 mM histidine on reabsorption of 2 mM arginine was studied in microperfusion experiments at initial pH values of 7.4 and 5.0. A conceptual difficulty arises here, however, because even at high buffer concentrations pH-values do not remain constant along the nephron [249, 312]. In any case, at the higher pH value where the imidazole amino group of histidine is largely unionized, no clear-cut inhibition of arginine reabsorption by histidine could be observed. However, inhibition at an initial pH of 5.0 was also non-significant, even though at that pH value the imidazole amino group is essentially completely ionized. Perhaps the relatively large imidazole ring prevents interaction of histidine with the transport system for diaminomonocarboxylic acids. This finding confirms the previously mentioned clearance results [138] which also failed to demonstrate interaction between L-histidine and L-arginine.

Normal plasma concentrations and clearance values for the other neutral L-amino acids *serine, alanine, threonine, valine,* and *methionine* are collected in Table 2. Little further information is available on the mechanism of their tubular reabsorption.

Reabsorption of filtered L-tyrosine exceeds 90%; in contrast, acetylated or methylated derivatives are absorbed far less readily [98]. Disubstituted L-tyrosine derivatives may even be secreted [181]. In addition, in clearance experiments a saturation of L-tyrosine reabsorption could be observed at high values of the filtered load. Just as L-tyrosine increases excretion of L-threonine, L-leucine,

Table 2. Clearance values and plasma concentrations of selected neutral amino acids in man [8, 79, 95, 107, 123, 289, 323]. Values for the other amino acids are given at the beginning of the appropriate chapters

Amino acid	Plasma concentration (mM)	Endogenous clearance (ml/min/1.73 m^2)
L-Serine	0.08–0.18	1.9–3.1
L-Alanine	0.23–0.62	0.3–1.0
L-Threonine	0.11–0.17	0.8–1.5
L-Tyrosine	0.04–0.10	1.0–3.0
L-Valine	0.16–0.31	0.1–0.3
L-Methionine	0.01–0.05	1.0–3.5

L-isoleucine and L-histidine [174] so excretion of L-tyrosine itself is stimulated by L-leucine and L-methionine [347]. An inhibitory action of maleate has not been clearly demonstrated [257]. In single tubules an approximate concentration profile under free-flow conditions has recently been reported [355] and found to resemble that of other neutral amino acids.

L-methionine is essentially completely reabsorbed up to a plasma level of 8 mM [111, 368]. In higher concentrations however it depresses transport of most amino acids [30, 59, 174, 347]. A possibly competitive inhibition of transport could here not be clearly distinguished from the toxic action of this amino acid [162, 218].

Renal handling of L-alanine is related in several ways to that of glycine and of AIB [171, 244, 280]. This would be expected on the basis of similarities in their chemical structure (Fig. 6). What is surprising is that infusion of L-alanine depresses relatively strongly reabsorption of almost all other amino acids [347]. The simple molecular structure of L-alanine, which so to speak forms part of the structure of all other amino acids, can however not account for the inhibition of their reabsorption. Indeed a similar effect should in that case be exerted by glycine; this however is not the case. Like other amino acids L-alanine is mostly reabsorbed in the proximal tubule [43, 354]. Osmotic diuresis reduces reabsorption of DL-alanine [149]. In its strong sensitivity to maleate, reabsorption of L-alanine resembles that of threonine, serine, valine and other amino acids [257].

Special consideration must be given to the synthetic neutral amino acid 1-amino-cyclopentane-carboxylic acid (*cycloleucine*). In this compound the α-carbon atom forms part of a five membered ring. Unlike in the case of proline the α-amino group remains outside the ring (see Fig. 6) so that the α-carbon atom is not optically active. In contrast to almost all other naturally occuring amino acids cycloleucine does therefore not possess two stereoisomeric forms.

Cycloleucine is transported across membranes but is not metabolized [4, 53, 54, 199]. As in the case of α-amino isobutyric acid it may serve therefore as a model compound for transport studies. The biological half-life of injected cycloleucine in plasma exceeds three weeks. Free filtrability in the kidney may

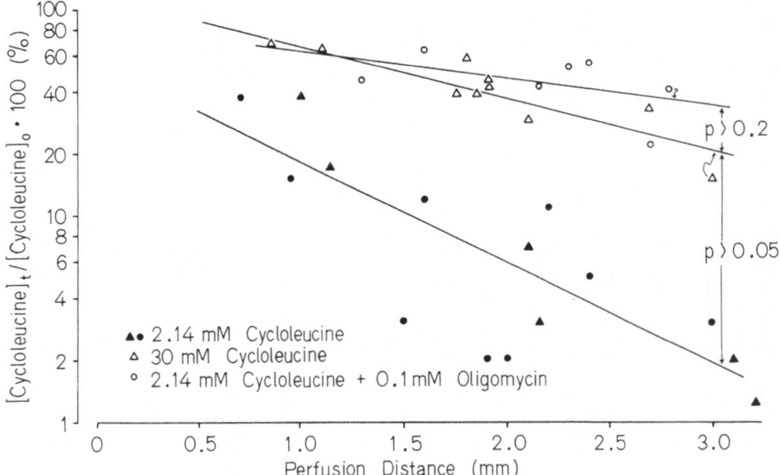

Fig. 20. Reabsorption of cycloleucine (Q=20nl/min). Oligomycin was added to the perfusion fluid with bovine albumin and ethanol (acc. to [165]). ● =cycloleucine (2.14 mM); ▲ =cycloleucine (2.14 mM)+albumin+ethanol (from [310])

be assumed so that reabsorption of the filtered load must amount to essentially 100%. In addition this amino acid is also secreted in the colon [53, 54]. In its small clearance cycloleucine resembles most of the other naturally occuring amino acids. What is striking however is the increased clearance of both neutral and dibasic amino acids observed following injection of cycloleucine [44, 154]. The inhibition of reabsorption of dibasic amino acids by a neutral compound like cycloleucine has recently been investigated in greater detail with microperfusion techniques [308, 310]. At an initial concentration of 2.14 mM cycloleucine moves out of the tubule faster than most other amino acids (see Fig. 20). At a distance of 3 mm over 98% of initial load have been reabsorbed. This high rate of fractional reabsorption is diminished (a) upon increasing the initial concentration to 30 mM, or (b) by addition of 0.1 mM oligomycin. We may plausibly conclude therefore that tubular reabsorption of cycloleucine is saturable and that it depends directly or indirectly on oxidative phosphorylation and/or sodium-potassium-ATPase, both of which can be inhibited by oligomycin.

The question also arises whether cycloleucine affects reabsorption of other amino acids. For this purpose reabsorption of L-arginine, glycine or L-phenylalanine was measured before and after addition of cycloleucine [310]. In all three cases a highly significant inhibition was observed. This result confirms observations made in earlier clearance studies [44, 154].

Provided arginine, phenylalanine, glycine and cycloleucine possess similar affinities for a hypothetical common transport system, the first three amino acids should in turn inhibit reabsorption of cycloleucine. This was found to be the case, however only with phenylalanine. In the case of glycine the inhibition is questionable whereas L-arginine appears unable to affect significantly reabsorption of cycloleucine ([310], see Fig. 21). These results argue against the concept

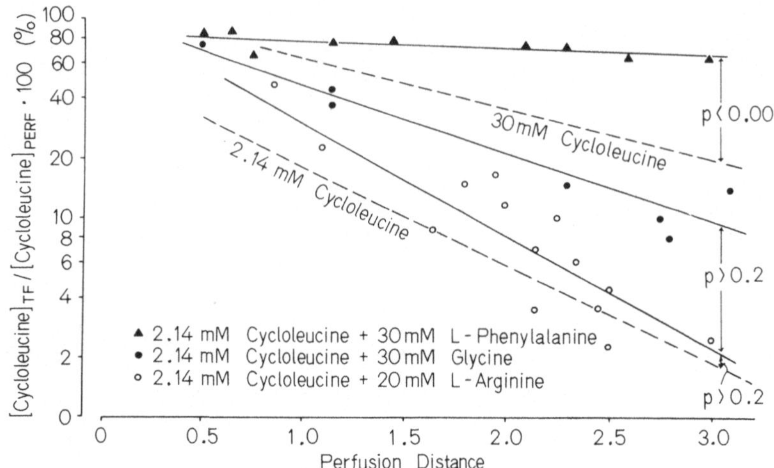

Fig. 21. Inhibition of cycloleucine reabsorption. The dashed lines were taken from Fig. 20 as controls. (Q=20nl/min) (from [310])

of a common transport system for arginine and cycloleucine. On the other hand phenylalanine and cycloleucine may well use a common transport mechanism in the proximal tubule.

As just mentioned L-phenylalanine strongly inhibits reabsorption of cycloleucine. If we ignore the unlikely possibility that phenylalanine decreases the passive permeability of tubular epithelium for cycloleucine, a maximal permeability coefficient (P) for cycloleucine can be calculated from its rate of fractional reabsorption in the presence of inhibitory concentrations of phenylalanine as 6×10^{-5} cm/sec (see Fig. 21). On the basis of the further preliminary assumption that reabsorption of cycloleucine can be described with simple Michaelis-Menten kinetics, the permeability coefficient just calculated permits calculation of the following transport constants for cycloleucine: $V_{max} = 36 \times 10^{-10}$ mol/cm²/sec; $K_m = 5.8$ mM [310].

Very recently tubular reabsorption of cycloleucine was investigated with the split-drop technique by LINGARD and collaborators [204]. These studies yielded a quotient V_{max}/P of about 25 mM. Using the figure of 6×10^{-5} cm/sec for the permeability coefficient, one obtains for V_{max} a value no greater than 15×10^{-10} mol/cm²/sec; this represents 40% of the value estimated above. Reasons for the slight discrepancy may be found in the use of different techniques. The discrepancy could also signify however (1) that the assumption of Michaelis-Menten kinetics is untenable, or (2) that different permeability coefficients determine movement of cycloleucine in the direction blood to lumen and lumen to blood, and finally also (3) that L-phenylalanine can affect passive permeation of cycloleucine.

Just as in the case of histidine [201] LINGARD et al. also found for cycloleucine an affinity constant (K_m) increasing by a factor of 10 along the proximal convoluted tubule. The average K_m value was 5.7 mM, in good agreement with that reported by SILBERNAGL (see above). All these values are calculated on the assumption that cycloleucine is reabsorbed by only one transport system. Especially because this assumption has been questioned [204] further investigation of the whole problem is required.

In summary, many instances have been reported of interaction between various neutral amino acids. This fact must not be equated however with adequate understanding of the whole area: we do not yet know (a) how many systems mediate reabsorption of neutral amino acids, or (b) which amino acids use what system. The present review can therefore summarize only facts as they appear at the moment and consider their preliminary significance (see Fig. 31).

B. Anionic Amino Acids

Transport processes for the acidic amino acids *L-glutamate* and *L-aspartate* and their respective amides, glutamine and asparagine, are difficult to investigate because of their rapid intracellular metabolism. In contrast to most other amino acids a great fraction of these substances is indeed metabolized by renal parenchyma [2, 3, 36, 80, 140, 149, 208, 246]. The following results must therefore be considered within the limitations imposed by metabolic alterations.

In stop-flow experiments a major site of reabsorption of glutamic and aspartic acids could be identified as the proximal tubule [43, 136, 267]. Micropuncture results confirmed this finding [13, 22, 355]. In semiquantitative micropuncture experiments 14% of the filtered load of glutamic acid were still found at the end of the proximal tubule. This fraction was reduced to a small portion of 1% in the final urine [355]. Microinjection into the distal tubule led to complete recovery of glutamic acid in urine [13]. Again the pars recta and possibly other portions of the loops of Henle must therefore be involved in reabsorption of glutamic acid.

The strong interaction between glutamic and aspartic acids during the process of their reabsorption has been known from clearance studies [138, 174, 244, 257, 347, 348]. The same interaction was also found in microperfusion experiments [13]. Although earlier experiments suggested a common transport system for glycine, alanine, arginine and glutamic acid [257], more recent investigations [13, 22, 138, 347, 348] emphasize specifically interaction between the transport of glutamic and aspartic acids. The suggestion of a common specific reabsorption mechanism for the two monoaminodicarboxylic acids is therefore well documented. It appears to be an active transport system [13], located on the luminal cell membranes [120].

C. Cationic Amino Acids

Because of their pathophysiological significance in cystinuria the tubular transport of basic amino acids (diaminomonocarboxylic acids: arginine, lysine and ornithine) has long been a focus of special interest: These three compounds are handled by the kidney in an almost identical fashion and are thus grouped together in the ALO-group.

Table 3. Clearance and plasma levels of endogenous L-arginine, L-lysine, and L-ornithine in man [8, 79, 107, 323, 325]

	L-arginine	L-lysine	L-ornithine
Clearance (ml/min/1.73 m^2)	0–6.6	0.2–1.9	0
Plasma level (mM)	0.019–0.14	0.09–0.34	0.05–0.1

The first clearance measurements for arginine and lysine were published already thirty years ago [30, 174, 208, 244, 368]. After the introduction of chromatographic techniques, the earlier values could be confirmed and defined more precisely [8, 79, 107, 323, 325]. Normal values are collected in Table 3. Higher clearances may result from overflow aminoaciduria in hyperargininemia or hyperlysinemia.

The site of reabsorption of the ALO-group in the nephron was identified as the proximal tubule [22, 43, 136, 139, 267]. Distal convoluted tubule and collecting duct are impermeable to L-arginine and L-lysine [22]. For these two compounds a clearly defined maximum rate of tubular reabsorption could be demonstrated [22, 137, 138, 208, 244, 368].

Reabsorption of L-arginine has also been studied in microperfusion experiments [297, 298, 300, 302, 303]. In a first series of such studies the concentration of L-arginine in the perfusion fluid was varied. In each case the decrease of radioactivity along the tubule was measured. At the lowest arginine concentration (0.36 mM) arginine was rapidly reabsorbed; increasing the initial concentration of arginine to 22 mM progressively decreased the fractional reabsorption rate down to only 10% over a tubular length of 3 mm (see Fig. 22).

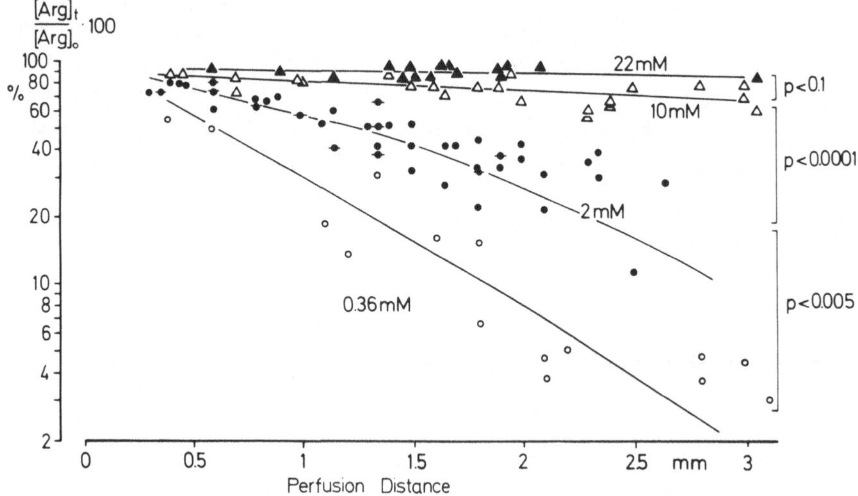

Fig. 22. Reabsorption of L-arginine at different initial concentrations (Q=20nl/min). The lines are calculated from Michaelis-Menten kinetics as further discussed in the text (from [300])

Unidirectional reabsorption of L-arginine in the proximal tubule is thus clearly saturable and also provides at least an approximate measure for net reabsorption. Indeed, the small rate of fractional reabsorption at 22 mM arginine shows that a contribution of diffusion to the overall reabsorption process can at best be only minimal. The low passive permeability of the proximal tubule for L-arginine can also be deduced from microinjection experiments which sought evidence for a possible secretion of arginine [24]. The maximum reabsorption rate (V_{max}) and the affinity constant (K_m) were estimated from the data of Fig. 22 using two different methods [300, 303]: For the first method 3 different transformations of the Michaelis-Menten equation were used:

$$\frac{1}{V_t}=\frac{1}{V_{max}}+\frac{K_m}{V_{max}}\cdot\frac{1}{C_t}; \qquad V_t=V_{max}-K_m\cdot\frac{V_t}{C_t}; \quad \frac{C_t}{V_t}=\frac{K_m}{V_{max}}+\frac{1}{V_{max}}\cdot\frac{1}{C_t}.$$

The rate of reabsorption (V_t) was calculated for each value depicted in Fig. 22 according to the equation

$$V_t=\frac{dC}{dt}\cdot\frac{r}{2}.$$

Here "C_t" stands for concentrations of arginine after a contact time "t" and "r" is the tubular radius. The approximation was made that over short time intervals concentration decreases at an exponential rate. Substitution of the value calculated for dC/dt on this basis yields the expression

$$V_t=\frac{r}{2}\cdot C_0\cdot a\cdot e^{at}$$

where C_0 is the initial concentration of L-arginine and "a" represents the slope of the semilogarithmic regression line relating C to t (i.e. the rate constant of the process).

To illustrate the saturation characteristics of arginine reabsorption Fig. 23 shows rate of reabsorption per tubular surface area (V) and per tubular length (V') as functions of concentration. Putting in each case the observed concentration C_t and the corresponding rate V_t into the three linear derivatives of the Michaelis-Menten equation one obtains significant linear correlations. The maximal reabsorption rate V_{max} and that arginine concentration at which the rate of reabsorption is half maximal, i.e. the K_m value, are shown in Table 4.

Table 4. Kinetic parameters of L-arginine reabsorption at $Q=20$ nl/min; values are based on the three linear transformations of the Michaelis-Menten equation and on its integrated form (HENRI plot [94, 168])

	$1/V$ vs. $1/C$	C/V vs. C	V vs. V/C	HENRI plot
V_{max} (10^{-10} mol·cm^{-2}·sec^{-1})	6.33	7.27	6.98	7.7
V'_{max} (10^{-12} mol·cm^{-1}·sec^{-1})	4.06	4.66	4.47	4.9
K_m (10^{-6} mol·cm^{-3})	0.99	1.48	0.92	1.12

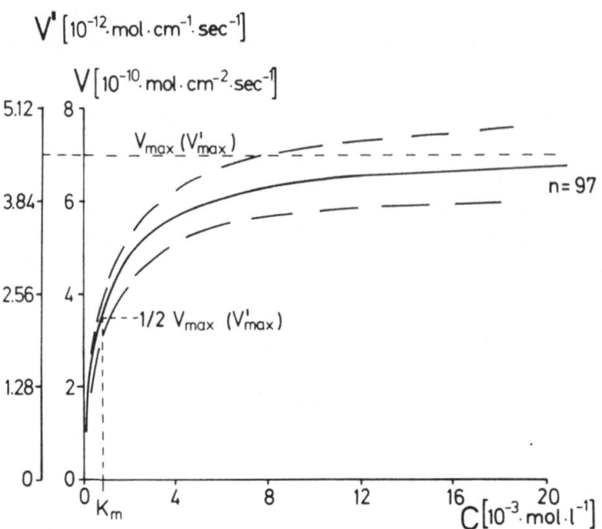

Fig. 23. Dependence of tubular reabsorption rate of L-arginine on concentration. Values are calculated on the basis of Fig. 22. The dotted lines represent one standard deviation from the estimate of the mean value for V or V' at a given concentration C (from [300])

For calculation of V the tubule is usually considered to be a cylinder. Comparison of results obtained at different flow rates suffers from unpredictable changes in surface area of the brush border [238]. It is therefore preferable to refer the rates of reabsorption only to tubular length at a given rate of tubular flow.

The second method used here for estimation of transport parameters was introduced by HENRI [94, 168]. Following this method one can calculate V_{max} and also K_m by substituting C_t and t directly in the integrated Michaelis-Menten equation. Values thus obtained for V_{max}, V'_{max} and K_m are also shown in Table 4.

The four procedures in Table 4 yield for L-arginine at a rate of tubular flow of 20 nl/min a maximal reabsorption rate (V_{max}) of 7.1×10^{-10} mol/cm^2/sec, a V'_{max} of 4.5×10^{-12} mol/cm/sec and K_m of approx. 1.2 mM. These values are now substituted in the Michaelis-Menten equation to permit integration and calculation of concentration as a function of time [300], indicated by the lines in Fig. 22.

The further question must now be raised whether transport parameters measured at 20 nl/min correspond to absolute values, or whether they will vary at different flow rates. To investigate this problem the tubular reabsorption of 2 mM L-arginine was studied at discrete perfusion rates ranging from 11 to 35 nl/min. The different perfusion rates, as expected, were associated with significant differences in rates of fractional reabsorption per unit length of perfused tubule.

These differences remain highly significant when the observed concentrations of arginine are plotted not against tubule length but against time of contact (Fig. 24). For instance, at a concentration C_t of 1.32 mM, comparison of transport rates calculated from corresponding contact times leads to the values of Table

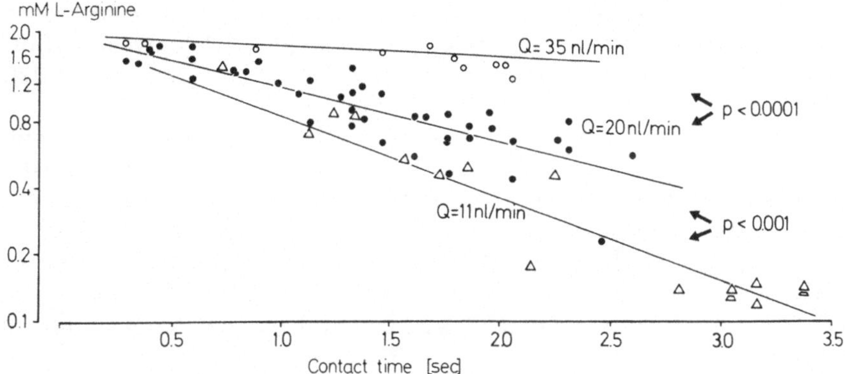

Fig. 24. Influence of rate of perfusion (Q) on L-arginine reabsorption (from [300])

5. Although no change in arginine reabsorption (V′) was seen between flow rates of 11 and 20 nl/min, at 35 nl/min V' was much decreased (Table 5). The relationship between rate of reabsorption and concentration shown in Fig. 23 is therefore valid only for a tubular flow rate of 20 nl/min and must be separately determined for any other rate of flow. The interval between 20 and 35 nl/min lies in the physiologically relevant range. Reabsorption rates may vary in that range both with the rate of glomerular filtration and with the extent of volume reabsorption along the proximal tubule. This fact could become especially important during osmotic diuresis.

Table 5. Reabsorption rate of L-arginine per tubular length (V') and per tubular surface (V) in proximal convolutions of the rat kidney at different flow rates (Q) but given arginine concentration C_t. r=tubular radius (according to [85]), t=contact time, C_0=initial arginine concentration

Q (nl/min)	r (10^{-4}cm)	C_0 (mM)	C_t (mM)	t (sec)	V (10^{-10}mol·cm^{-2}·sec^{-1})	V' (10^{-12}mol·cm^{-1}·sec^{-1})
11	8.1	2.0	1.32	0.44	5.05	2.57
20	10.2	2.0	1.32	0.70	4.00	2.56
35	11.7	2.0	1.32	2.40	1.33	0.97

It is not certain what causes these variations but two possibilities must be considered. In the first place reabsorption of ionized L-arginine could depend on the transtubular electrical potential difference. The latter, in turn, might well be a function of tubular flow rate. Secondly one might assume that at a high perfusion rate a radial concentration gradient is established in the tubular lumen. This would lead to a higher arginine concentration in the center of the lumen than nearer the tubular epithelium. For any given average concentration across the whole tubule such an event would thus lead to underestimation of the kinetic parameters of reabsorption as of course the rate of this reabsorption depends only on the L-arginine concentration next to the tubular wall.

The effect of some inhibitors of cell metabolism on L-arginine reabsorption was investigated by microperfusion experiments [306]. Oligomycin (0.1 mM), cyanide (5 mM) and carbonyl–cyanide–m–chlorophenyl hydrazone (1.0 mM) are able to inhibit L-arginine reabsorption in the proximal tubule.

The discussion so far has been restricted to the reabsorption of L-arginine. There is little doubt however that L-lysine and L-ornithine share at least one common reabsorption mechanism with L-arginine. As previously mentioned, many examples have been observed of these amino acids competing with one another for transport. The system, excluding as it does all other naturally occuring amino acids, clearly represents a highly specific mechanism. The question then arises of the precise structural prerequisites permitting interaction with this transport system.

Using the clearance technique it has been shown recently [205] that N^{ε}-monomethyllysine is reabsorbed like lysine, whereas the di- and trimethylated derivatives are excreted by the kidney practically like inulin. In microperfusion experiments [298, 303] the effect on L-arginine reabsorption of a series of L-arginine derivatives including L-lysine and L-ornithine was investigated. The derivatives (see formulae in Fig. 25) were added in a concentration of 20 mM. The results previously obtained with 2 mM and 22 mM arginine (Fig. 22) served as controls.

L-lysine and L-ornithine strongly depressed L-arginine transport, so that its fractional reabsorption decreased from around 80% to approx. 35% at a tubular length of 3 mm (Fig. 26a). This diminution was almost as great as that seen when the concentration of arginine was increased from 2 to 22 mM.

Elongation of the carbon chain of L-arginine yields L-homoarginine, an amino acid capable of strongly inhibiting reabsorption of L-arginine (see Fig. 26a) an interaction also observed in cystinuric patients [66]. This effect parallels the undiminished inhibitory action of L-lysine compared with that of L-ornithine. L-2,3-diaminopropionic acid with its much shorter carbon chain on the other hand influences the reabsorption of L-arginine to a much smaller degree (Fig. 26b). This reduced inhibition could of course relate to a second cause: the 3-amino group of diaminopropionic acid has a pK value of 6.73 [317]. Unlike the case of ornithine and lysine at a physiological pH, this second amino group will be only partially ionized. A similar consideration applies also to the imidazole group of L-histidine as previously mentioned.

D-arginine exhibits a much smaller affinity for the reabsorptive system than does its stereoisomer L-arginine (Fig. 26b). 5-aminovaleric acid, obtained by deamination of L-ornithine, is a far weaker inhibitor than is its parent substance (Fig. 26b). Decarboxylated arginine (agmatine) on the other hand continues strongly to depress L-arginine reabsorption (Fig. 26a).

At the other end of the arginine molecule, substitution of the 5-methylene group by an oxygen atom (L-canavanine) does not affect inhibitory potency (Fig. 26a). In contrast, changing the guanidine residue of L-arginine into a urea group (L-citrulline) completely destroys the inhibitory property of the molecule (Fig. 26c). The urea group is no longer ionizable. The assumption thus seems justified that ionizability of the second amino group is required before a molecule can react with the mechanism responsible for reabsorbing the ALO-group. This assumption is strengthened by the observation that L-homoserine which is dis-

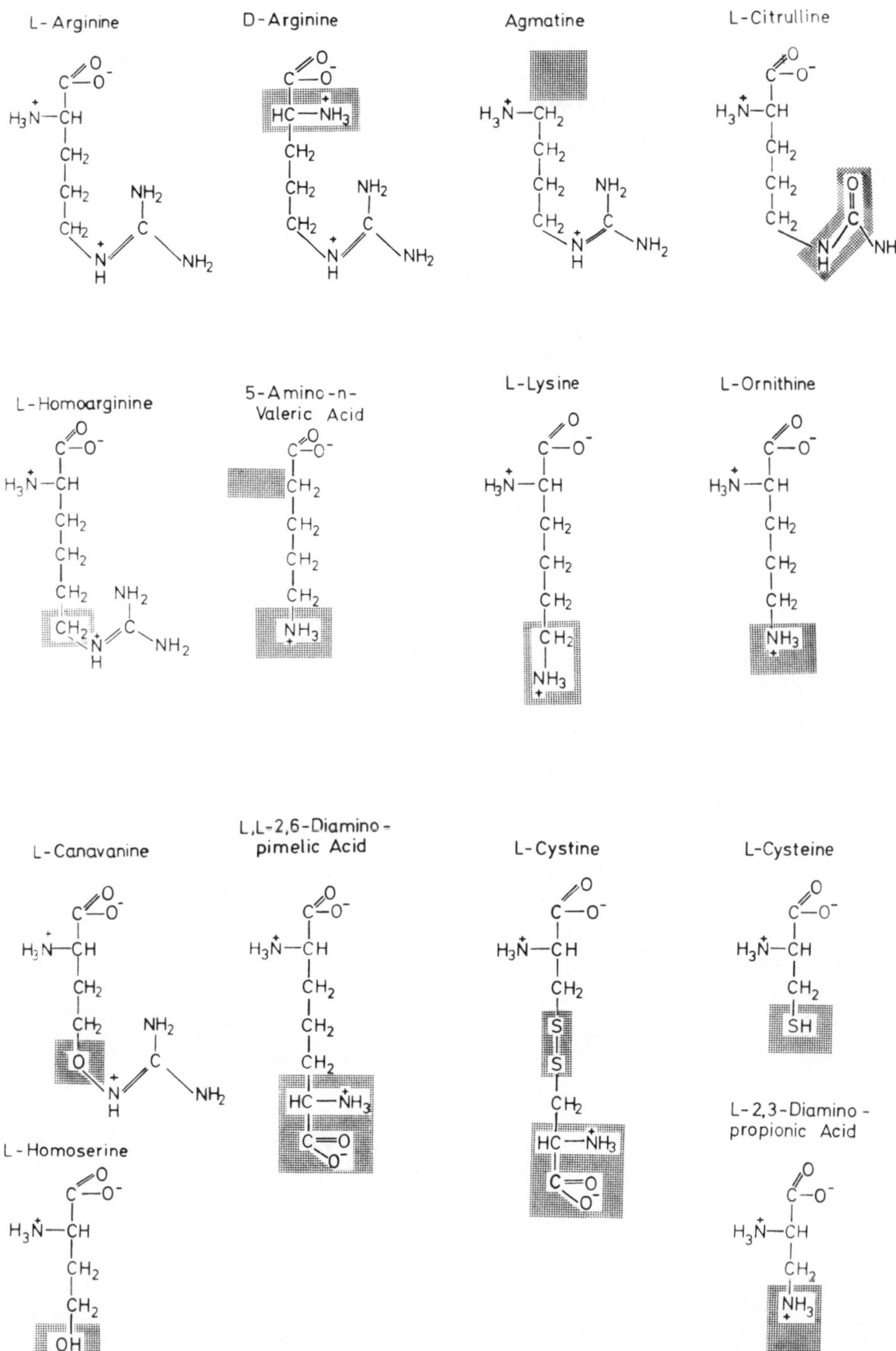

Fig. 25. Chemical structures of L-arginine derivatives mentioned in this paper. The structural differences from L-arginine are marked by shaded areas (from [303])

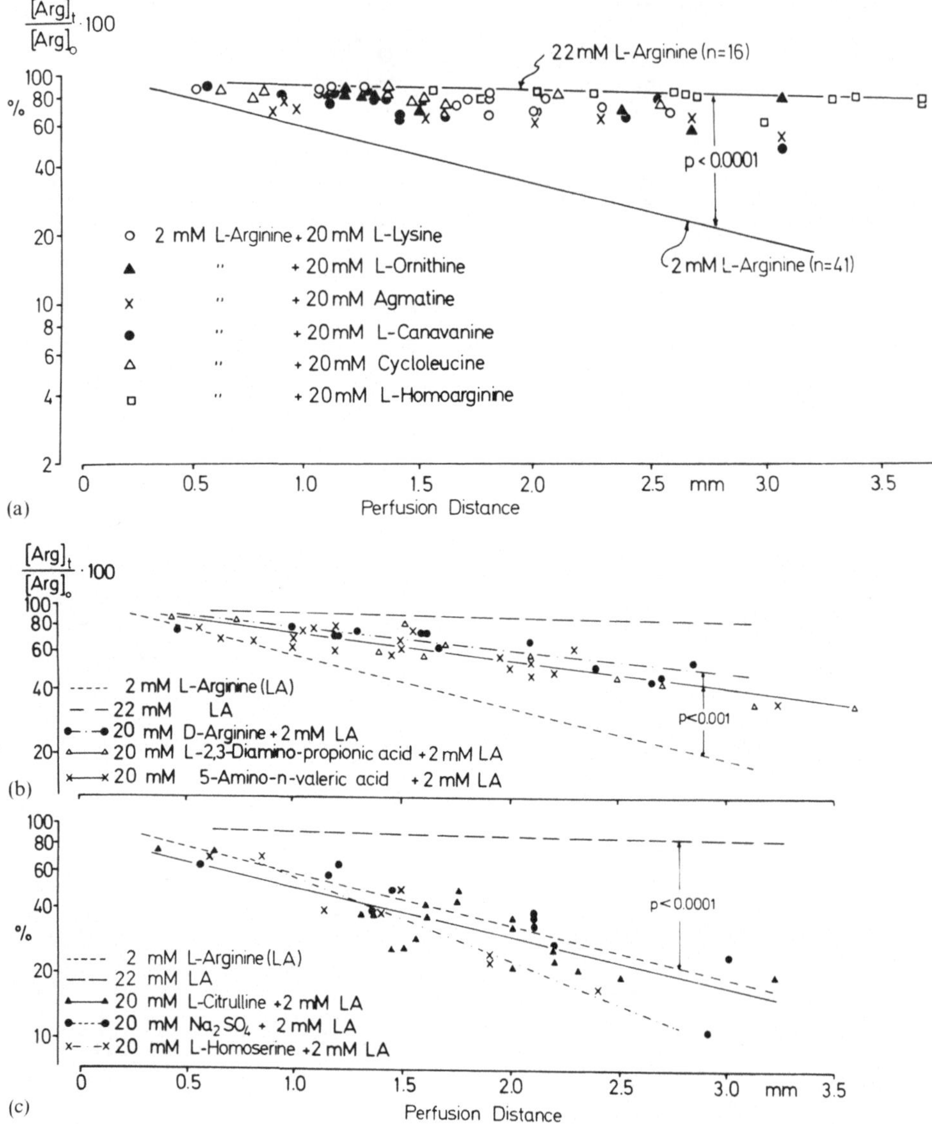

Fig. 26a–c. Effect of various L-arginine derivatives (see Fig. 25) on L-arginine reabsorption. (Q = 20 nl/min). Control lines for L-arginine alone are taken from Fig. 22 (from [303])

tinguished from L-ornithine by the substitution of a hydroxyl for the 5-amino group, does not inhibit arginine reabsorption at all (Fig. 26c). The possible binding of the OH-group of homoserine to the transport receptor with the aid of a sodium ion, suggested on the basis of *in vitro* experiments with ascites tumor cells [56, 57], does not appear to be germane to renal transport. Nor can a sulfhydryl group replace the ω-amino configuration. Indeed L-cysteine in contrast to L-2,3-diaminopropionic acid does not inhibit the reabsorption of L-arginine (see Fig. 13). Similarly 2,6-diaminopimelic acid is inactive in this system (Fig. 13). This compound is distinguished from L-ornithine by the introduction of a carboxyl group on C 5. Diaminodicarboxylic acids thus are unable to react with the tubular reabsorptive system for the ALO-group. This finding is of importance in considering tubular transport of cystine, another diaminodicarboxylic acid, and was discussed in greater detail in Section VI.A.3.

In summary therefore substances can react with the tubular reabsorption system for the ALO-group if they possess the following formula:

$$^{-}OOC-\underset{\underset{+}{NH_3}}{\overset{H}{\overset{|}{\underset{|}{C}}}}-(CH_2)_x-\underset{+}{\overset{R_1}{\overset{|}{N}}}H-R_2$$

The optimal value of x is 3 to 4. The methylene group adjacent to the nitrogen can be replaced by an oxygen atom. The radicals R_1 and R_2 must permit ionization of the vicinal nitrogen. Such a condition is met by a guanido residue as well as by the two hydrogen atoms of a primary amino group; a urea residue will not serve this purpose. Although the amino group in the imidazole ring of L-histidine can also be ionized, the compound exerts only minimal inhibition on arginine reabsorption. The relatively large volume of the imidazole ring may here be of significance as previously discussed.

The amino acid must be present in the L-form. The carboxy group however seems not to be essential for inhibitory activity. It must be emphasized that reaction with the transport system does not necessarily imply transport of the inhibitor itself. The following experiments bear further on this point.

Of the several derivatives of L-arginine discussed above L-lysine and L-canavanine were chosen for more detailed study. As a first experiment their own tubular reabsorption was determined in presence and absence of L-arginine. Results of these experiments are collected in Fig. 27. Here it is seen that tubular

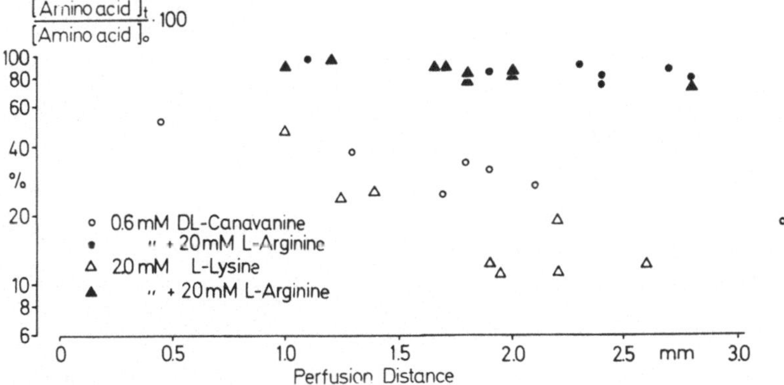

Fig. 27. Reabsorption of L-lysine and DL-canavanine

reabsorption both of 2 mM L-lysine and of 0.6 mM DL-canavanine is inhibited almost completely by addition of 20 mM L-arginine. The effect of lysine is not particularly surprising in view of the generally accepted common reabsorptive system for members of the ALO-group. The behaviour of canavanine could not be predicted however. It suggests competition with L-arginine for the reabsorptive mechanism rather than a possible non-specific toxic action [212, 287]. In the long run of course such a competition could also become toxic for the cell, especially in relation to protein metabolism.

It is also important to quantify the mutual effects on transport of L-arginine on the one hand and of L-canavanine and L-lysine on the other by determination of the reabsorptive and inhibitory constants. For this purpose rates of fractional reabsorption of L-arginine were determined at different initial concentrations (from 0.36 to 6.0 mM) and constant inhibitor concentration (4 mM). Results are shown in Figs. 28 and 29. Transforming in each case the length of the

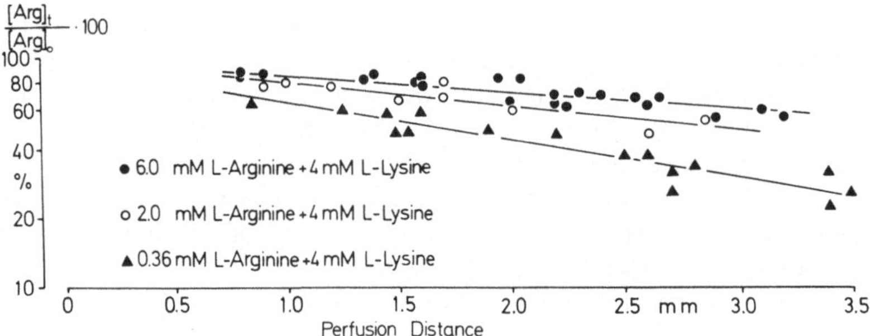

Fig. 28. L-arginine reabsorption at different initial L-arginine but constant L-lysine concentrations (from [303])

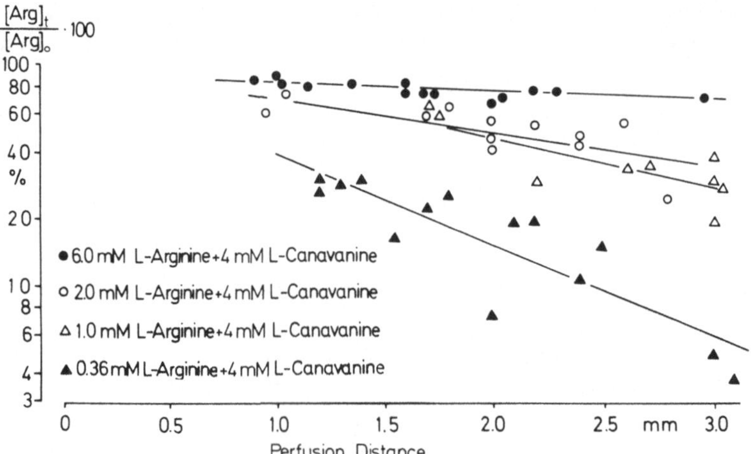

Fig. 29. L-arginine reabsorption at different initial L-arginine but constant L-canavanine concentrations (from [303])

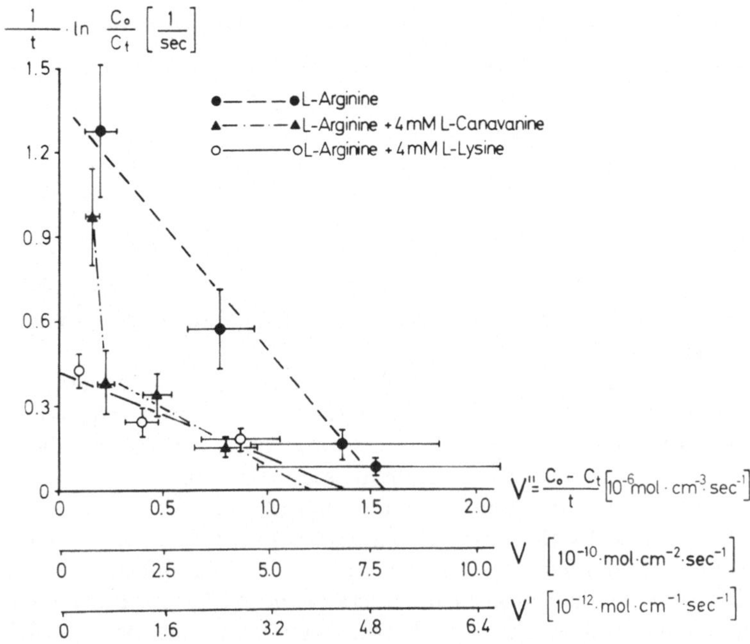

Fig. 30. Henri plot [94, 168] of L-arginine reabsorption in presence of inhibitors (from [303])

perfused tubule into contact time and substituting the corresponding absolute concentration C_t in the formula of HENRI [94, 168] leads to Fig. 30. Controls were derived from data in Fig. 22.

Results obtained with L-lysine are represented by an approximately straight line which clearly differs from the controls in its slope ($-1/K_m$), but not significantly in the x intercept (V''_{max}). Competition between L-lysine and L-arginine for a common transport system can thus readily be demonstrated on the basis of Michaelis-Menten kinetics.

In presence of L-canavanine a similar picture is obtained only if the 14 values obtained at the lowest arginine concentration (0.36 mM) are rejected. These values, in contrast to those obtained at higher concentrations of L-arginine, cannot be significantly distinguished from controls. It follows that in the case of arginine concentrations below approx. 1 mM L-canavanine no longer inhibits reabsorption of the ALO-group.

These results can be interpreted in terms of 2 different tubular reabsorptive systems for L-arginine. One of these with its constants was described previously and can be competitively inhibited by L-canavanine. The second system whose existence must be postulated is much less sensitive to canavanine and would appear to be already saturated at an arginine concentration of 1 mM.

From the data on inhibition of arginine reabsorption by L-lysine and L-canavanine the apparent affinity constant K_P for L-arginine in presence of the inhibitor, and the affinity constant K_i of the inhibitor itself, as well as the values

Table 6. Kinetic parameters of L-arginine reabsorption in presence and absence of L-lysine and L-canavanine. Values are based on results shown in Figs. 22, 28, and 29

	K_m (mM)	K_p (mM)	K_i (mM)	V_{max} (10^{-10}mol·cm^{-2}·sec^{-1})	V'_{max} (10^{-12}mol·cm^{-1}·sec^{-1})
L-Arginine (controls)	1.12	—	—	7.7	4.9
+4 mM L-Lysine	—	3.3	2.0	6.9	4.4
+4 mM L-Canavanine	—	2.4	3.6	6.0	3.8

for V_{max} can be calculated. Values obtained are collected together with the control constants in Table 6.

The affinity of L-canavanine for the ALO-reabsorption system (K_i=3.6 mM) is smaller than that of L-lysine (K_i=2.0 mM) and L-arginine (K_m=1.12 mM). True competition between L-arginine and L-canavanine can however be deduced from the V_{max} value of 6.0×10^{-10} mol/cm^2/sec found in the presence of canavanine at initial L-arginine concentrations higher than 1 mM. The nature of the inhibition exerted by the other derivatives of L-arginine mentioned remains to be determined. Equally unsolved is the reason why L-alanine and L-methionine [347] can influence reabsorption of members of the ALO-group. Interaction between arginine and cycloleucine was further discussed in Section VI.A.4.

The tubular transport system of the ALO-group has been studied in greater detail than that of the other amino acids. Nevertheless little is known about the mechanism itself. The data obtained in the experiments described here do define however a series of caracteristics of this transport system and should thus contribute to the elucidation of its mechanism of action.

VII. Summary and Conclusions

The aim of this review was in the first place that of summarizing our knowledge of amino acid transport in the kidney. The scheme in Fig. 31 attempts to describe in outline the present understanding of these processes. As shown in the figure, a series of transport mechanisms must be postulated to explain the handling of the various groups of amino acids. In many cases a specific aminoaciduria results from malfunction of one of these systems.

The extensive knowledge now available on specificity, localization, kinetics and inhibition of renal amino acid transport system remains in strong contrast to our ignorance of the molecular mechanisms involved. Among recent attempts to explain transport in biochemical terms special mention must be made of the γ-glutamyl cycle of ORLOWSKI and MEISTER [236]. Although this scheme cannot fully explain all transport phenomena such as e.g. that of imino acids, it nevertheless offers an interesting and stimulating possible molecular model. The continued elucidation of receptor mechanisms and of the coupling of transport to metabolic energy are perhaps the major problem for future research in this area.

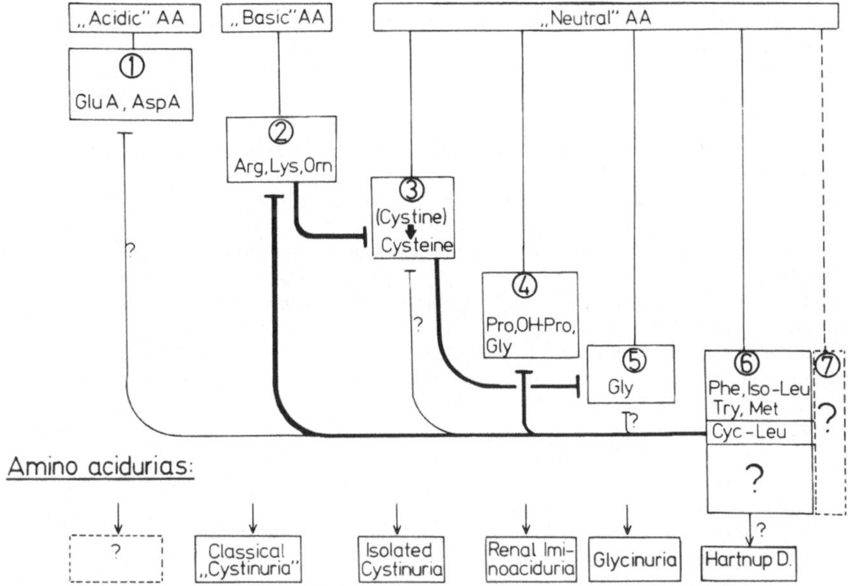

Fig. 31. Tubular reabsorption systems for L-amino acids (from [311])

Further refinements of micropuncture techniques will permit analysis of many questions at present unanswered. In addition a continued and possibly more important role is likely to be played by studies of solute transport and binding by membrane vesicles and components. Such work has already provided interesting insights into problems of sugar transport [189], and its application to amino acids has also been reported [343]. These methods promise to be useful also for the study of amino acid receptors, and for the separation of luminal from peritubular transport systems [167].

Recent electrophysiological studies [131] and investigation of the coupling between amino acid and electrolyte transport [342] represent an additional useful approach. Finally the dissection of amino acid transport by study of its fetal and neonatal development, or by genetic analysis of metabolic diseases is also likely to lead to further useful insights. No single technique, however, can provide answers to more than a portion of the important questions remaining in the field of renal amino acid transport.

References

1. Ackermann, D., Kutscher, F.: Über das Vorkommen von Lysin im Harn bei Cystinurie. Z. Biol. **57**, (bzw. **39**), 355 (1912).
2. Addae, S.K., Lotspeich, D.W.: Relation between glutamine utilization and production in metabolic acidosis. Amer. J. Physiol. **215**, 269 (1968).
3. Addae, S.K., Lotspeich, D.W.: Glutamine balance in metabolic acidosis as studied with the artificial kidney. Amer. J. Physiol. **215**, 278 (1968).

4. AKEDO, H., CHRISTENSEN, H.N.: Transfer of amino acids across the intestine: a new model amino acid. J. biol. Chem. **237**, 113–117 (1962).
4a. AMSTRONG, W., NUNN, A.S. (Ed.): Intestinal transport of electrolytes, amino acids, and sugars. Springfield-Illinois: C.C. Thomas (1971).
5. ANGIELSKI, S., ROGULSKI, J.: Acta biochim. pol. **6**, 411 (1959) quoted from [257].
6. ANGIELSKI, S., NIEMIRO, R., MAKAREWIEZ, W., ROGULSKI, J.: Acta biochim. pol. **5**, 431 (1958) quoted from [257].
7. ANGIELSKI, S., ROGULSKI, J., MADONSKA, L.: Acta biochim. pol. **7**, 269 (1960) quoted from [257].
8. ARROW, V.K., WESTALL, R.G.: Amino acid clearance in cystinuria. J. Physiol. (Lond.) **142**, 141 (1958).
9. ASATOOR, A.M., LACEY, B.W., LONDON, D.R., MILNE, M.D.: Amino acid metabolism in cystinuria. Clin. Sci. **23**, 285 (1962).
10. AUERBACH, V.H., DI GEORGE, A.M., BALDRIDGE, R.C., TOURTELOTTE, C.D., BRIGHAM, M.P.: Histidinemia. A deficiency of histidase resulting in the urinary excretion of histidine and of imidazolepyrovic acid. J. Pediat. **60**, 487 (1962).
11. AUSIELLO, D.A., SEGAL, S., THIER, S.O.: Cellular accumulation of L-Lysine in rat kidney cortex in vivo. Amer. J. Physiol. **222**, 1472 (1972).
12. AYER, J.L., SCHIESS, W.A., PITTS, R.F.: Independence of phosphat reabsorption and glomerular filtration in the dog. Amer. J. Physiol. **151**, 168 (1947).
13. BAINES, DEWITT, A., MOREL, F.: Absorption of acidic amino acid from proximal tubule fluid. Proceedings of the IVth International Congress of Nephrology, p. 293, Stockholm, (1969).
14. BANK, H., CRISPIN, M., EHRLICH, D., SZEINBERG, A.: Iminoglycinuria. A defect of renal tubular transport. Israel J. med. Sci. **8**, 606 (1972).
15. BANK, N., AYNEDJIAN, H.S.: Techniques of microperfusion of renal tubules and capillaries. Yale J. Biol. Med. **45**, 312 (1972).
16. BARON, D.N., DENT, C.E., HARRIS, H., HART, E.W., JEPSON, J.B.: Hereditary pellagra-like skin rash with temporary cerebellar ataxia, constant aminoaciduria and other bizarre biochemical features. Lancet **1956 II**, 421.
17. BARTSOCAS, C.S., THIER, S.O., CRAWFORD, J.D.: Transport of L-methionine in rat intestine and kidney. Pediat. Res. **8**, 673 (1974).
18. BARTTER, F.C., LOTZ, M., THIER, S.O., ROSENBERG, L.G., POTTS, J.T.JR.: Cystinuria. Ann. intern. Med. **62**, 796 (1965).
19. BERGER, H.: Die Amino-Stickstoff-Ausscheidung im Harn in Abhängigkeit vom Lebensalter. Ann. paediat. (Basel) **186**, 338 (1956).
20. BERGER, H.: Hereditäre chronische Hyperaminoacidurien. Mod. Probl. Pädiat. **3**, 238 (1957).
21. BERGER, E.A., HEPPEL, L.A.: A binding protein involved in the transport of cystine and diaminopimelic acid in escherichia coli. J. biol. Chem. **247**, 7684 (1972).
22. BERGERON, M., MOREL, F.: Amino acid transport in rat renal tubules. Amer. J. Physiol. **216**, 1139 (1969).
23. BERGERON, M.: Renal amino acid accumulation in maleate-treated rats. Rev. Can. Biol. **30**, 267 (1971).
24. BERGERON, M., VADEBONCOER, M.: Antiluminal transport of L-Leucine following microinjections in peritubular capillaries of the rat. Nephron **8**, 355 (1971).
25. BERGERON, M., VADEBONCOER, M.: Microinjections of L-Leucine into tubules and peritubular capillaries of the rat. II. The maleic acid model. Nephron **8**, 367 (1971).
26. BERGLUND, F., LOTSPEICH, W.D.: Effect of various amino acids on the renal tubular reabsorption of inorganic sulphate in the dog. Amer. J. Physiol. **185**, 539 (1956).
27. BERLINER, R.W., KENNEDY, T.J., HILTON, J.G.: Effect of maleic acid on renal function. Proc. Soc. exp. Biol. (N.Y.) **75**, 791–794 (1950).
28. BERZELIUS, J.J.: Calculus urinairs. Traite Chem. **7**, 424 (1833) quoted from [335].
29. BEYER, K.H., WRIGHT, L.D., RUSSO, H.F., SKEGGS, H.R., PATCH, E.A.: The renal clearance of essential amino acids: tryptophan, leucine, isoleucine and valine. Amer. J. Physiol. **146**, 330 (1946).
30. BEYER, K.H., WRIGHT, L.D., SKEGGS, H.R., RUSSO, H.F., SHANER, G.A.: Renal clearance of essential amino acids: their competition for reabsorption by the renal tubules. Amer. J. Physiol. **151**, 202 (1947).

31. BICKEL, H., BAAR, H.S., ASTLEY, R., DOUGLAD, A., FINCH, E., HARRIS, H., HARVEY, C.C., HICKMANS, E.M., PHILPOTT, M.G., SMALLWOOD, W.C., SMELLIE, J.M., TEALL, C.G.: Cystine storage disease with amino-aciduria and Dwarfism (Lignac-Fanconi disease). Acta Paediat. **42**, 1 (1952).
32. BICKEL, H.: Die Entwicklung der biochemischen Läsion bei der Lignac-Fanconischen Krankheit. Helv. paediat. Acta **10**, 259 (1955).
33. BICKEL, H.: Neuere Erkenntnisse zur hepatocerebralen Degeneration (Wilsonsche Krankheit). Mod. Probl. Pädiat. **3**, 215 (1957).
34. BLIX, G.: Über die Löslichkeitsverhältnisse von Cystin im Harn. Hoppe-Seylers Z. physiol. Chem. **178**, 109 (1928).
35. BOJESEN, E., LEYSSAK, P.P.: The kidney cortex slice technique as a model for sodium transport in vivo. Acta Physiol. scand. **65**, 20 (1965).
36. BOURKE, E., FINE, A., SCOTT, J.: Mechanism of ammoniagenesis in human kidney. Biochem. J. **125**, 94P (1971).
37. BRAND, E., HARRIS, M.M., BILOON, S.: Cystinuria. The excretion of a cystine complex which decomposes in the urine with the liberation of free cystine. J. biol. Chem. **86**, 315 (1930).
38. BRAND, E., CAHILL, G.F., HARRIS, M.M.: Cystinuria. II. The metabolism of cystine, cysteine, methionine, and glutathione. J. biol. Chem. **109**, 69 (1935).
39. BRAND, E., CAHILL, G.F.: Canine cystinuria. III. J. biol. Chem. **114**, XV. (1936).
40. BRAND, E., CAHILL, G.F., KASSELL, B.: Canine cystinuria. V. Family history of two cystinuric Irish terriers and cystine determination in dog urine. J. biol. Chem. **133**, 431 (1940).
41. BRIGHAM et al.: J. clin. Invest. **39**, 1633 (1960), quoted from [92a].
42. BRODEHL, J., GILISSEN, K., KOWALEWSKI, S.: Isolierter Defekt der tubulären Cystin-Rückresorption in einer Familie mit idiopathischem Hypoparathyroidismus. Klin. Wschr. **45**, 38 (1967).
43. BROWN, J.L., SAMIY, A.H., PITTS, R.F.: Localisation of aminonitrogen reabsorption in the nephron of the dogs. Amer. J. Physiol. **200**, 370 (1961).
44. BROWN R.R.: Aminoaciduria resulting from cycloleucine administration in man. Science **157**, 432–434 (1967).
45. BURG, M.B., ORLOFF, J.: Oxygen consumption and active transport in separated renal tubules. Amer. J. Physiol. **203**, 327 (1962).
46. CAIN, A.R.R., HOLTON, J.B.: Histidinaemia: a child and his family. Arch. Dis. Childh. **43**, 62 (1968).
47. CARTON, D., DHONDT, F., DE SCHRIJVER, F., SAMYN, W., KINT, J., DELBEKE, M.J., HOOFT, C.: Histidinemia. Helv. paediat. Acta **2**, 127 (1970).
48. CHAN, Y.L., HUANG, K.C.: Microperfusion studies on renal tubular transport of tryptophan derivatives in rats. Amer. J. Physiol. **221**, 575 (1971).
49. CHAN, Y.L., HUANG, K.C.: Renal excretion of D-tryptophan, 5-hydroxytryptamine, and 5-hydroxyindoleacetic acid in rats. Amer. J. Physiol. **224**, 140 (1973).
50. CHILDS, B., NYHAN, W.L., BORDEN, M., BARD, L., COOKE, R.E.: Idiopathic hyperglycinemia and hyperglycinuria: a new disorder of amino acid metabolism. I. Pediatrics **27**, 522 (1961).
51. CHINARD, F.P., DELEA, A.C.: Luminal and anti-luminal transport characteristics of certain amino acids. Proceedings of the IVth International Congress of Nephrology, p. 294, Stockholm, (1969).
52. CHRISTENSEN, H.N., RIGGS, T.R.: Structural evidences for chelation and Schiff's base formation in amino acid transfer into cells. J. biol. Chem. **220**, 265–278 (1956).
53. CHRISTENSEN, H.N., CLIFFORD, J.B.: Excretion of l-aminocyclopentanecarboxylic acid in man and the rat. Biochim. biophys. Acta (Amst.) **62**, 160 (1962).
54. CHRISTENSEN, H.N., JONES, J.C.: Amino acid transport models: renal reabsorption and resistance to metabolic attack. J. biol. Chem. **237**, 1203 (1962).
55. CHRISTENSEN, H.N.: Some transport lessons taught by the organic solute, in: Biological transport. Perspect. Biol. Med. **10**, 471 (1967).
56. CHRISTENSEN, H.N., HANDLOGTEN, M.E., THOMAS, E.L.: Proc. nat. Acad. Sci. (Wash.) **63**, 948 (1969).
57. CHRISTENSEN, H.N.: Electrolyte effects on the transport of cationic amino acids. In: Na-linked transport of organic solutes. (HEINZ, E., ed.), p. 39, Berlin-Heidelberg-New York: Springer 1972.
58. CHRISTOPHEL, W., DEETJEN, P.: Mikroperfusionsuntersuchungen zum tubulären Transport von Glycin. Pflügers Arch. ges. Physiol. **297**, R52 (1967).

59. CHRISTOPHEL, W.: Der tubuläre Transport der Aminosäure Glycin: Mikroperfusionsuntersuchungen am proximalen Konvolut der Rattenniere in vivo et situ. Dissertation, München (1969).
60. CHRISTOPHEL, W., CHAN, Y.L., WILLIAMS, W.M., HUANG, K.C.: Microperfusion and stop-flow studies on renal tubular transport of phenylalanine. Fed. Proc. **31**, (1972).
61. CLARANCE, G.A., BOWMAN, J.K.: Further case of histidinemia. Brit. med. J. **1**, 1019 (1966).
62. COHEN, J.J., WHITMAN, E.: Renal utilization and excretion of α-ketoglutarate in dog: effect of alkalosis. Amer. J. Physiol. **204**, 795 (1963).
63. COLINDRES, R.E., LECHENE, C.: Technical problems associated with collection of distal tubular fluid in the rat. Yale J. Biol. Med. **45**, 233 (1972).
64. CORNELIUS, C.E., BISHOP, J.A., SCHAFFER, M.H.: A quantitative study of amino aciduria in Dachshunds with a history of cystine urolithiasis. Cornell Vet. **177**, 177 (1967).
65. CORNER, B.D., HOLTON, J.B., NORMAN, R.M., WILLIAMS, P.M.: A case of histidinemia controlled with a low histidine diet. Pediatrics **41**, 1074 (1968).
66. COX, B.D., CAMERON, J.S.: Homoarginine in cystinuria. Clin. Sci. **46**, 173–182 (1974).
67. CRANE, C.W., TURNER, A.W.: Amino acid patterns of urine and blood plasma in a cystinuric Labrador dog. Nature (Lond.) **177**, 237 (1956).
68. CRAWHALL, J.C., SCOWEN, E.F.: Effect of penicillamine on cystinuria. Brit. med. J. **I**, 588 (1963).
69. CRAWHALL, J.C., SCOWEN, E.F., WATTS, R.W.: Further observation on use of D-penicillamine in cystinuria. Brit. med. J. **1**, 1411 (1964).
70. CRAWHALL, J.C., SEGAL, S.: Sulphocysteine in the urine of the blotched kenya genet. Nature (Lond.) **208**, 1320 (1965).
71. CRAWHALL, J.C., THOMSON, C.J.: Renal secretion of cystine in cystinuria. J. clin. Invest. **44**, 1038 (1965).
72. CRAWHALL, J.C., THOMSON, C.J.: Cystinuria: effect of D-penicillamine on plasma and urinary cystine concentrations. Science **147**, 1459 (1965).
73. CRAWHALL, J.C., SEGAL, S.: Dithiothreitol in the study of cysteine transport. Biochim. biophys. Acta (Amst.) **121**, 215 (1966).
74. CRAWHALL, J.C., SEGAL, S.: The intracellular cysteine/cystine ratio in kidney cortex. Biochem. J. **99**, 1965 (1966).
75. CRAWHALL, J.C., SCOWEN, E.F., THOMPSON, C.J., WATTS, R.W.E.: The renal clearance of amino acids in cystinuria. J. clin. Invest. **46**, 1162 (1967).
76. CRAWHALL, J.C., SEGAL, S.: The intracellular ratio of cysteine and cystine in various tissues. Biochem. J. **105**, 891 (1967).
77. CRAWHALL, J.C., WATTS, R.W.E.: Cystinuria. Amer. J. Med. **45**, 736 (1968).
78. CROSS, R.J., TAGGART, J.V.: Renal tubular transport: accumulation of PAH by rabbit kidney slices. Amer. J. Physiol. **161**, 181 (1950).
79. CUSWORTH, D.C., DENT, C.E.: Renal clearance of amino acids in normal adults and in patients with aminoaciduria. Biochem. J. **74**, 550 (1960).
80. DAMIAN, A.C., PITTS, R.F.: Rates of glutaminase I and glutamine synthetase in rat kidney in vivo. Amer. J. Physiol. **218**, 1249 (1970).
81. DATTA, S.P., HARRIS, H.: Urinary amino-acid patterns of some mammals. Ann. Eugen. (Lond.) **18**, 107 (1953).
82. DAVIDMAN, M., LALONE, R.C., ALEXANDER, E.A., LEVINSKY, N.G.: Some micropuncture techniques in the rat. Amer. J. Physiol. **221**, 1110 (1971).
83. DAVIES, H.E., ROBINSON, M.J.: A case of histidinemia. Arch. Dis. Childh. **43**, 62 (1968).
84. DEBRÉ, R., MARIE, J., CLERET, F., MESSIMY, R.: Rachitisme tardif coexistant avec une nephrite chronique et une glycosurie. Arch. Méd. Enf. **37**, 597 (1934).
85. DEETJEN, P., BOYLAN, J.W.: Glucose reabsorption in the rat kidney: microperfusion studies. Pflügers Arch. ges. Physiol. **299**, 19 (1968).
86. DEETJEN, P., SILBERNAGL, S.: Some new developments in continuous microperfusion technique. Yale J. Biol. Med. **45**, 301 (1972).
87. DENT, C.E., ROSE, G.A.: Amino acid metabolism in cystinuria. Abstr. Commun. 1st Int. Congr. Biochem. Cambridge (1949).
88. DENT, C.E., ROSE, G.A.: Amino acid metabolism in cystinuria. Quart. J. Med. **20**, 205 (1951).
89. DENT, C.E., SENIOR, B., WALSHE, J.M.: The pathogenesis of cystinuria. II. Polarographic studies of the metabolism of sulphur-containing amino-acids. J. clin. Invest. **33**, 1216 (1954).
90. DENT, C.E., SENIOR, B.: Studies on the treatment of cystinuria. Brit. J. Urol. **27**, 317 (1955).

91. DENT, C.E.: Klassifikation der Aminoacidurie. Scand. J. clin. Lab. Invest. Suppl. **13**, 21 (1957).
92. DE TONI: Remarks upon the relation between renal rickets (renal dwarfism) and renal diabetes. Acta paediat. (Uppsala) **16**, 479 (1933).
92a. DIEM, K., LENTNER, C. (GEIGY A.G.): Documenta Geigy, Wissenschaftliche Tabellen. 7. Auflage, Basel, J.R. Geigy A.G. (1968).
93. DIES, F., SANDOVAL, G., MARTINEZ, R., GARZA, R., ORDONEZ, A.: Effects of probenecid, alkalosis and glycine on net renal uptake and on tubular reabsorption of lactate in dogs. Rev. Invest. clin. **26**, 111–123 (1974).
94. DIXON, M., WEBB, E.C.: Enzymes. 2nd ed., p. 114, London: Longmans 1964.
95. DOOLAN, P.D., HARPER, H.A., HUTCHIN, M.E., SHREEVE, W.W.: Renal clearance of eighteen individual amino acids in human subjects. J. clin. Invest. **34**, 1247 (1955).
96. DOOLAN, P.D., HARPER, H.A., HUTCHIN, M.E., ALPEN, E.L.: The renal tubular response to amino acid loading. J. clin. Invest. **35**, 888 (1956).
97. DOOLAN, P.D., HARPER, H.A., HUTCHIN, M.E., ALPEN, E.L.: Renal clearance of lysine in cystinuria. Amer. J. Med. **23**, 416 (1957).
98. DOTY, J.R.: Reabsorption of certain amino acids and derivatives by the kidney tubules. Proc. Soc. exp. Biol. (N.Y.) **46**, 129 (1941).
99. DRUMMOND, K.N., MICHAEL, A.F.: Specificity of the inhibition by certain amino acids. Nature (Lond.) **201**, 1333 (1964).
100. EATON, A.G., FERGUSON, F.P., BYER, F.T.: The renal reabsorption of amino acids in dogs: valine, leucine, and isoleucine. Amer. J. Physiol. **145**, 491 (1946).
101. EFRON, M.L.: Treatment of hydroxyprolinemia and hyperprolinemia. Amer. J. Dis. Child **113**, 166 (1967).
102. EFRON, M.L.: Familial hyperprolinemia. Report of a second case, associated with congenital renal malformations, hereditary hematuria and mild mental retardation, with demonstration of an enzyme defect. New. Engl. J. Med. **272**, 1243 (1965).
103. EFRON, M.L.: Aminoaciduria. New. Engl. J. Med. **272**, 1058 (1965).
104. EMERY, F.A., GOLDIC, L., STERN, J.: Hyperprolinemia: clinical and biochemical family study. J. ment. Defic. Res. **12**, 187 (1968).
105. ETTINGER, B., KOLB, F.O.: Factors involved on crystal formation in cystinuria. In vivo and in vitro crystallization dynamics and a simple, quantitative colorimetric assay for cystine. J. Urol. (Baltimore) **106**, 106 (1971).
106. EVERED, D.F.: Species differences in amino acid excretion by mammals. Comp. Biochem. Physiol. **23**, 163 (1967).
107. EVERED, D.F.: The excretion of amino acids by the human: a quantitative study with ion-exchange chromatography. Biochem. J. **62**, 416 (1956).
108. FANCONI, G.: Die nicht-diabetischen Glykosurien und Hyperglykämien des älteren Kindes. Jb. Kinderheilk. **133**, 17 (1931).
109. FANCONI, G.: Der nephrotisch-glykosurische Zwergwuchs mit hypophosphatämischer Rachitis. Jb. Kinderheilk. **147**, 199 (1936).
110. FANCONI, G.: Der nephrotisch-glykosurische Zwergwuchs mit hypophosphatämischer Rachitis. Dtsch. med. Wschr. **62**, 1169 (1936).
111. FERGUSON, F.P., EATON, A.G., ASHMAN, J.S.: Renal reabsorption of methionine in normal dogs. Proc. Soc. exp. Biol. (N.Y.) **66**, 582 (1947).
112. FOULKES, E.C., MILLER, B.F.: Steps in PAH transport by kidney slices. Amer. J. Physiol. **196**, 86 (1959).
113. FOULKES, E.C., FORSTER, R.P.: Potassium transport by kidney slices of lophius americanus. Bull. Mt. Des. Island Biol. Lab. **4**, 44 (1959).
114. FOULKES, E.C., PAINE, C.M.: The uptake of monocarboxylic acid by rat diaphragm. J. biol. Chem. **236**, 1019 (1961).
115. FOULKES, E.C.: Effects of heavy metals on renal aspartate transport and the nature of solute movement in kidney cortex slices. Biochim. biophys. Acta (Amst.) **241**, 815 (1971).
116. FOULKES, E.C.: Renal aspartate transport. Fred. Proc. **31**, (1972).
117. FOULKES, E.C., GIESKE, T.: Specificity and metal sensitivity of renal amino acid transport. Biochim. biophys. Acta (Amst.) **318**, 439 (1973).
118. FOULKES, E.C.: Site of the functional lesion responsible for amino aciduria after administration of organo mercurials and other metal compounds. In: Mercury, Mercurials and Mercaptans,

4th Rochester International Conference on Environmental Toxicity. (MILLER, M.W., CLARKSON, T.W., eds), Springfield: C.C. Thomas 1973.
119. FOULKES, E.C.: Peritubular transport of urate and amino acids in rat kidney. In: Amino acid transport (ed. SILBERNAGL, S.) and uric acid (ed. LANG, F., GREGER, R.). Stuttgart: Thieme 1975 (in press).
120. FOULKES, E.C.: Cellular localisation of amino acid carriers in renal tubules. Proc. Soc. exp. Biol. (N.Y.) **139**, 1032 (1972).
121. FOWLER, D.J., HARRIS, H., WARREN, F.: Plasma cystine levels in cystinuria. Lancet **1952 I**, 544.
122. FOX, M., THIER, S., ROSENBERG, L.E., KISER, W., SEGAL, S.: Evidence against a single renal transport defect in cystinuria. New. Engl. J. Med. **270**, 556 (1964).
123. FRAME, E.G.: The levels of individual free amino acids in plasma of normal man at various intervals after a high protein meal. J. clin. Invest. **37**, 1719 (1958).
124. FRASER, G.R., FRIEDMAN, A.I., PATTON, V.M., WADE, D.N., WOOLF, L.I.: Iminoglycinuria. A "harmless" inborn error of metabolism? Hum. Genet. **6**, 362 (1968).
125. FREEDMAN, B.S., YOUNG, J.A.: Microperfusion study of L-histidine transport by the rat nephron. Austr. J. exp. Biol. med. Sci. **47**, 10 (1969).
126. FRIEDMANN, E.: Der Kreislauf des Schwefels in der organischen Natur. Ergebn. Physiol. **1**, 15 (1902).
127. FRIMPTER, G.W., HORWITH, M., FURTH, E., FELLOWS, R.E., THOMPSON, D.D.: Inulin and endogenous amino acid renal clearance in cystinuria: evidence for tubular secretion. J. clin. Invest. **41**, 281 (1962).
128. FRIMPTER, G.W.: Cystinuria: metabolism of the disulfide of cysteine and homocysteine. J. clin. Invest. **42**, 1956 (1963).
129. FRIMPTER, G.W.: Cystinuria: intravenous administration of (S 35) cystine and (S 35) cysteine. Clin. Sci. **31**, 207 (1966).
130. FRÖMTER, E., LÜER, K.: Free flow pontential profile along rat proximal tubule. Pflügers Arch. **339**, R47 (1973).
131. FRÖMTER, E., RUMRICH, G., ULLRICH, K.J.: Phenomenologic description of Na^+, Cl^- and HCO_3^- absorption from proximal tubules of the rat kidney. Pflügers Arch. **343**, 189–220 (1973).
132. FRÖMTER, E., GESSNER, K.: Active transport potentials, membrane diffusion potentials and streaming potentials across rat kidney proximal tubules. Pflügers Arch. **351**, 85–98 (1974).
133. GAITONDE, M.K., GAULL, G.: A procedure for the quantitative analysis of the sulphur amino acids of the rat tissue. Biochem. J. **102**, 959 (1967).
134. GARROD, A.: Inborn errors of metabolism. Lancet **1908 II**, 1.
135. GASSER, G., PREISINGER, A.: Cystinsteine. Klin. Wschr. **38**, 1130 (1960).
136. GAYER, J., GEROK, W.: Die Lokalisierung der L-Aminosäuren-Rückresorption in der Niere durch Stop-flow-Analysen. Klin. Wschr. **39**, 1054 (1961).
137. GEROK, W., GAYER, J.: Investigation on the reabsorption of amino acids in the kidney of the dog. Proc. 1st. Int. Congr. Nephrol. (Evian) p. 720 (1961).
138. GEROK, W., GAYER, J.: Die tubuläre Rückresorption der L-Aminosäuren in der Niere des Hundes. Transportmaxima und competitive Hemmung. Klin. Wschr. **39**, 540 (1961).
139. GEROK, W., GAYER, J.: Investigation on the reabsorption of L-amino acids in the kidney of the dog with special reference to diamino-monocarbonic acids. Excerpta Medica, Intern. Congr. **29**, 572 (1964).
140. GEROK, W., NIETH, H.: Untersuchungen über die renale arteriovenöse Aminosäuredifferenz am Menschen. In: Normale und pathologische Funktion des Nierentubulus. 3. Symp. Dtsch. Ges. Nephrol., Berlin (1964), Bern, Huber (1965).
141. GEROK, W.: Primäre Tubulopathien. Stuttgart: Thieme 1969.
142. GEROK, W.: Defekte renaltubulärer Transportsysteme für Aminosäuren. Med. Klin. **70**, 301–312 (1975).
143. GERTZ, K.H.: Transtubuläre Natriumchloridflüsse und Permeabilität für Nichtelektrolyte im proximalen und distalen Konvolut der Rattenniere. Pflügers Arch. ges. Physiol. **267**, 336 (1963).
144. GERTZ, K.H.: Stationary perfusions methods. Yale J. Biol. Med. **45**, 265 (1972).
145. GHADIMI, H., PARTINGTON, M.W., HUNTER, A.: A familial disturbance of histidine metabolism. New. Engl. J. Med. **265**, 221 (1961).
146. GIESKE, T., FOULKES, E.C.: Acute effects of cadmium on proximal tubular function in rabbit. Toxicol. appl. Pharmacol. **27**, 292 (1974).

147. GILLISSEN, J., TAUGNER, R.: Die Nierenausscheidung von Ascorbinsäure, Glykokoll und Alanin bei der Katze. Z. ges. exp. Med. **134**, 179 (1961).
148. GOETTSCH, E., LYTTLE, J.D., GRIM, W.M., DUNBAR, P.: The renal amino acid clearance in the normal dog. Amer. J. Physiol. **140**, 688 (1944).
149. GOLDSTEIN, L.: Pathways of glutamine deamination and their control in the rat kidney. Amer. J. Physiol. **213**, 983 (1967).
150. GOOD, N.E., WINGET, G.D., WINTER, W., CONNOLLY, T.N., IZAWA, S., SINGH, R.M.: Hydrogen ion buffer for biological research. Biochemistry (Wash.) **5**, 467 (1966).
151. GOODMAN, S., MCINTYRE, C.A., O'BRIEN, D.: Impaired intestinal transport of proline in a patient with familial iminoaciduria. J. Pediat. **71**, 246 (1967).
152. GOULDEN, B.E., LEAVER, J.L.: Low voltage electrophoresis as a screening test for the diagnosis of canine cystinuria. Vet. Rec. **80**, 244 (1967).
153. GOYER, R.A., REYNOLDS, J.JR., BURKE, J., BURKHOLDER: Hereditary renal disease with neurosensory hearing loss, prolinuria and ichtyosis. Amer. J. med. Sci. **256**, 166 (1968).
154. GOYER, R.A., REYNOLDS, J.O., ELSTON, R.C.: Characteristics of the aminoaciduria resulting from cycloleucine administration in pair-fed rats. Proc. Soc. exp. Biol. (N.Y.) **130**, 860 (1969).
155. GREEN, D.F., MORRIS, M.L., CAHILL, G.F., BRAND, E.: Canine cystinuria. II. Analysis of cystine calculi and sulfur distribution in the urine. J. biol. Chem. **114**, 91 (1936).
156. GREENSTEIN, J.P., WINITZ, M.: Chemistry of the amino acids, vol. I, p. 486. New York-London-Sydney: J. Wiley and Sons 1961.
157. GUTMAN, A.B., YÜ, T.F.: Bull. N.Y. Acad. Med. **34**, 287 (1958).
158. GYÖRY, A.Z.: Reexamination of the split oil droplet method as applied to kidney tubules. Pflügers Arch. **324**, 328 (1971).
159. GYÖRY, A.Z., KINNE, R.: Energy source for transepithelial sodium transport in rat renal proximal tubules. Pflügers Arch. **327**, 234 (1971).
160. GYÖRY, A.Z.: Sources of error in and limitation in the use of t1/2 as a measure of tubular reabsorptive capacity. Yale J. Biol. Med. **45**, 269 (1972).
161. HAGIHIRA, H., LIN, E.C.C., SAMIY, A.H., WILSON, T.H.: Active transport of lysine, ornithine, arginine and cystine by the intestine. Biochem. biophys. Res. Commun. **4**, 478 (1961).
162. HARDWICK, D.F., APPLEGARTH, D.A., COCKCROFT, D.M., ROSS, P.M., CALDER, R.J.: Pathogenesis of methionine-induced toxicity. Metabolism **19**, 381 (1970).
163. HARRIS, H., MITTWOCH, U., ROBSON, E., WARREN, F.L.: The pattern of amino-acid excretion in cystinuria. Ann. hum. Genet. **19**, 196 (1955).
164. HARRIS, H., MITTWOCH, U., ROBSON, E., WARREN, F.L.: Phenotypes and genotypes in cystinuria. Ann. hum. Genet. **20**, 57 (1955).
165. HARRIS, H., ROBSON, E.: Cystinuria. Amer. J. Med. **22**, 774 (1957).
166. HARRISON, H.E., HARRISON, H.C.: Experimental production of renal glycosuria, phosphaturia, and aminoaciduria by injection of maleic acid. Science **120**, 606 (1954).
167. HEIDRICH, H.G., KINNE, R., KINNE-SAFRAN, E., HANNING, K.: "The polarity of the proximal tubule cell in rat kidney". Different surface charges for the brush-border microvilli and plasma membranes from the basal infoldings. J. Cell Biol. **54**, 232 (1972).
168. HENRI, V.: Über das Gesetz der Wirkung des Invertins. Z. physik. Chem. **39**, 194–216 (1902).
169. HERBERT, J.D., COULSON, R.A., HERNANDEZ, T.: Free amino acid in the caiman and rat. Comp. Biochem. Physiol. **17**, 583 (1966).
170. HIERHOLZER, K., CADE, R., GURD, R., KESSLER, R., PITTS, R.F.: Stop-flow analysis of renal reabsorption and excretion of sulfate in the dog. Amer. J. Physiol. **198**, 833 (1960).
171. HILLMAN, R.E., ALBRECHT, I., ROSENBERG, L.E.: Identification and analysis of multiple glycine transport systems in isolated mammalian renal tubules. J. biol. Chem. **243**, 5566 (1968).
172. HILLMAN, R.E., ROSENBERG, L.E.: Amino acid transport by isolated mammalian renal tubules. II. Transport systems for L-proline. J. biol. Chem. **244**, 4494 (1969).
173. HILLMAN, R.E., ROSENBERG, L.E.: Binding and transport of proline by mammalian renal tubule cell preparations. Proceeding of the IVth International Congress of Nephrology, p. 294, Stockholm (1969).
174. HILLMAN, R.E., ROSENBERG, L.E.: Amino acid transport by isolated mammalian renal tubules. III. Binding of L-proline by proximal tubule membranes. Biochim. biophys. Acta (Amst.) **211**, 318 (1970).
175. HOLTON, J.B., LEWIS, F.J.W., MOORE, G.R.: Biochemical investigation of histidinemia. J. clin. Path. **17**, 671 (1964).

176. HOLTON, J.B.: The effect of the histidine load on plasma levels and renal clearances of other amino acids. Clin. chim. Acta **21**, 241 (1968).
177. HOLTZAPPLE, P.G., BOVES, C.F., REA, C.F., SEGAL, S.: Amino acid uptake by kidney and jejunal tissue from dogs with cystine stones. Science **166**, 1525 (1969).
178. HOLTZAPPLE, P.G., REA, C.F., GENEL, M., SEGAL, S.: Cycloleucine inhibition of amino acid transport in human and rat kidney cortex. J. Lab. clin. Med. **75**, 818 (1970).
179. HOLTZAPPLE, P.G., REA, C.F., BOVEE, K., SEGAL, S.: Characteristics of cystine and lysine transport in renal and jejunal tissue from cystinuric dogs. Metabolism **20**, 1016 (1971).
180. HOOFT, C., TIMMERMANS, J., SNOECK, J., ANTENER, I., OYAERT, W., VAN DEN HENDE, C.: Methionine malabsorption syndrome. Ann. Pédiat. **205**, 73 (1965).
181. HUANG, K.C.: Renal excretion of L-tyrosine and its derivatives. J. Pharmacol. exp. Ther. **134**, 257 (1961).
182. JOSEPH, R., RIBIERRE, M., JOB, J.C., GIRAULT, M.: Maladie familiale associante des convulsions a debut tres precoce, une hyperalbuminorachie et une hyperaminoacidurie. Arch. franç. Pédiat. **15**, 374 (1958) quoted from [16].
183. KAMIN, H., HANDLER, PH.: Effect of infusion of single amino acids upon excretion of other amino acids. Amer. J. Physiol. **164**, 654 (1951).
184. KÄSER, H., COTTIER, P., ANTENER, I.: Die Glukoglycinurie, ein neues familiäres Syndrom. Helv. paediat. Acta **16**, 586 (1961).
185. KASHGARIAN, M., STÖCKLE, H., GOTTSCHALK, C.W., ULLRICH, K.J.: Transtubular electrochemical potentials of sodium and chloride in proximal and distal renal tubules of rats during antidiuresis and water diuresis (Diabetes insipidus). Pflügers Arch. ges. Physiol. **277**, 89 (1963).
186. KEKOMÄKI, M., VISAKORPI, J.K., PERHEENTUPA, J., SAXEN, L.: Familial protein intolerance with deficient transport of basic amino acids. Acta paediat. (Uppsala) **56**, 617 (1967).
187. KEKOMÄKI, M., RAIHA, N.C.R., PERHEENTUPA, J.: Enzymes of urea synthesis in familial protein intolerance with deficient transport of basic amino acids. Acta paediat. (Uppsala) **56**, 631 (1967).
188. KING, J.S., WAINER, A.: Cystinuria with hyperuricemia and methioninuria. Biochemical study of a case. Amer. J. Med. **43**, 125 (1967).
189. KINNE, R., KINNE-SAFRAN, E., MURER, H.: Uptake of D-glucose by brush border microvilli and membranes from lateral and basal infoldings isolated from rat kidney cortex. Pflügers Arch. **343**, R46 (1973).
190. KIRK, E.: Studies on the amino acid clearance. Acta med. scand. **89**, 450 (1936).
191. KLEINZELLER, A., CORT, J.H.: The mechanism of action of mercurial preparation on transport processes and the role of thiol groups in the cell membrane of renal tubular cells. Biochem. J. **67**, 15 (1957).
192. KLEINZELLER, A., KOLINSKA, J., BENES, I.: Transport of glucose and galactose in kidney-cortex cells. Biochem. J. **104**, 843 (1967).
193. KOPELMANN, H., ASATOOR, A.M., MILNE, M.D.: Hyperprolinemia and hereditary nephritis. Lancet **1964 II**, 1075.
194. KRIZEK, V.: Zystinurie und Zystinsteinkrankheiten. Med. Klin. **62**, 1230 (1967).
195. LABELLE, W.C., MILLER, D.S., LERNER, J.: Interactions between leucine and arginine transport in chicken small intestine. Biochem. biophys. Res. Commun. **45**, 131 (1971).
196. LA DU, B.N.: Histidinemia. Amer. J. Child. **113**, 88 (1967).
197. LA DU, B.N., HOWEL, R.R., JACOBY, G.A., SEEGMILLER, J.E., SOBER, E.K., ZANNONI, V.G., CANBY, J.P., ZIEGLER, L.K.: Clinical and biochemical studies on two cases of histidinemia. Pediatrics **32**, 216 (1963).
198. LANG, F., GREGER, R., DEETJEN, P.: Handling of uric acid by the rat kidney. II. Microperfusion studies on bidirectional transport of uric acid in the proximal tubule. Pflügers Arch. **335**, 257 (1972).
199. LIANG, M., IRVING, J.L., WILSON, J.E.J.: Elisha Mitchel Sci. Soc. **81**, 25 (1963).
200. LIGNAC, G.O.E.: Über Störungen des Cystinstoffwechsels. Münch. med. Wschr. **71**, 1016 (1924).
201. LINGARD, J., RUMRICH, G., YOUNG, J.A.: Reabsorption of L-glutamine and L-histidine from various regions of the rat proximal convolution studied by stationary microperfusion: evidence that the proximal convolution is not homogeneous. Pflügers Arch. **342**, 1–12 (1973).
202. LINGARD, J., RUMRICH, G., YOUNG, J.A.: Kinetics of L-histidine transport in the proximal convolution of the rat nephron studied using the stationary microperfusion technique. Pflügers Arch. **342**, 13–28 (1973).

203. LINGARD, J.M., TURNER, B., WILLIAMS, D.B., YOUNG, J.A.: Endogenous amino acid clearance by the rat kidney. Austral. J. exp. Biol. med. Sci. **52**, 687–695 (1974).
204. LINGARD, J.M., GYÖRY, A.Z., YOUNG, J.A.: Inhomogeneity of cycloleucine reabsorption in the proximal convolution of the rat kidney. Pflügers Arch. (in press), (1975).
205. LÖWER, R., LANGE, H.W., HEMPEL, K.: Ausscheidung und renale Behandlung von N^{ε}-Monomethyl-Lysin, N^{ε}-Dimethyl-Lysin und N^{ε}-Trimethyl-Lysin. Nieren- und Hochdruckkrankheiten. **3**, XVII (1974).
206. LOEWY, A., NEUBERG, C.: Über Cystinurie. Hoppe-Seylers Z. physiol. Chem. **43**, 338 (1904).
207. LOWENSTEIN, L.M., SMITH, J., SEGAL, S.: Amino acid transport in the rat renal papilla. Biochim. biophys. Acta (Amst.) **150**, 73 (1968).
208. LOTSPEICH, W.D., PITTS, R.F.: The role of amino acids in the renal tubular secretion of ammonia. J. biol. Chem. **168**, 611 (1947).
209. MACKENSIE, S., SCRIVER, C.R.: Transport of L-proline and α-aminoisobutyric acid in the isolated rat kidney glomerulus. Biochim. biophys. Acta (Amst.) **241**, 725 (1971).
210. MCCARTHY, C.F., BORLAND, J.L., LYNCH, H.J., OWEN, E.E., TYOR, M.P.: Defective uptake of basic amino acids and L-cystine by intestinal mucosa of patients with cystinuria. J. clin. Invest. **43**, 1518 (1964).
211. MCLEOD, M.E., TYOR, M.P.: Transport of basic amino acids by hamster intestine. Amer. J. Physiol. **213**, 163 (1967).
212. MEISTER, A.: Biochemistry of the amino acids, 2nd ed., vol. I, p. 252. New York-London: Academic Press 1965.
213. MICHAEL, A.F., DRUMMOND, K.N.: Inhibitory effect of certain amino acids on renal tubular absorption of phosphate. Canad. J. Physiol. Pharmacol. **45**, 103 (1967).
214. MILNE, M.D., CRAWFORD, M.A., GIRAO, C.B., LOUGHRIDGE, L.W.: The metabolic disorder in hartnup disease. Quart. J. Med. **29**, 407 (1960).
215. MILNE, M.D., ASATOOR, A., LOUGHRIDGE, W.: Hartnup disease and cystinuria. Lancet **1961 I**, 51.
216. MILNE, M.D., ASATOOR, A.M., EDWARDS, K.D., LOUGHRIDGE, L.W.: The intestinal absorption defect in cystinuria. Gut **2**, 323 (1961).
217. MILNE, M.D.: Disorders of amino-acid transport. Brit. med. J. **1**, 327 (1964).
218. MILNE, M.D.: Pharmacology of amino acids. Clin. Pharmacol. Ther. **9**, 484 (1968).
219. MILNE, M.D.: Amino acid metabolism in cystinuria. Biochem. J. **122**, 9P (1971).
220. MINDER, F.C., DUBACH, U.C., ANTENER, I.: Hereditäre Nephropathie und Schwerhörigkeit (mit Aminosäuren- und Fettstoffwechselstörungen in einer Familie aus der Schweiz). Z. klin. Med. **158**, 601 (1965).
221. MOHYUDDIN, F., SCRIVER, C.R.: Amino acid transport in mammalian kidney: multiple systems for imino acids and glycine in rat kidney. Amer. J. Physiol. **219**, 1 (1970).
222. MOLLICA, F., PAVONE, L., ANTENER, I.: Pure familial hyperprolinemia: Isolated inborn error of aminoacid metabolism without other anomalies in a sicilian family. Pediatrics **48**, 225 (1971).
223. MORIN, C.L., THOMPSON, M.W.: Biochemical and genetic studies in cystinuria: Observations on double heterozygotes of genotype I/II. J. clin. Invest. **50**, 1961 (1971).
224. MORRIS, R.C., MC SHERRY, E., KRANHOLD, J.F., SEBASTIAN, A.: Modulation of proximal and distal tubule function in the Fanconi Syndrom. Birth defects **6**, 22 (1970).
225. MUDGE, G.H.: Electrolyte and water metabolism of rabbit kidney slices: Effect of metabolic inhibitors. Amer. J. Physiol. **167**, 206 (1951).
226. MUNCK, B.G.: Amino acid transport by the small intestine of the rat. Evidence against interactions between sugars and amino acids at the carrier level. Biochim. biophys. Acta (Amst.) **156**, 192 (1968).
227. MUNCK, B.G., SCHULTZ, S.G.: Interactions between leucine and lysine transport in rabbit ileum. Biochim. biophys. Acta (Amst.) **183**, 182 (1969).
228. MUNCK, B.G., SCHULTZ, S.G.: Lysine transport across isolated rabbit ileum. J. gen. Physiol. **53**, 157 (1969).
229. MUNCK, B.G.: Interactions between lysine, Na^+ and Cl^- transport in rat jejunum. Biochim. biophys. Acta (Amst.) **203**, 424 (1970).
230. MUNCK, B.G.: Interactions between sugar and amino acid transport in rat jejunum. Acta physiol. Scand. **82**, 32A (1971).
231. MURTHY, L., FOULKES, E.C.: Movement of solutes across luminal cell membranes in kidney tubules of the rabbit. Nature (Lond.) **213**, 180 (1967).

232. NECHAY, B.R., PALMER, R.F., CHINOY, D.A., POSEY, V.A.: The problem of $Na^+ - K^+$ adenosine triphosphatase as the receptor for diuretic action of mercurials and ethacrynic acid. J. Pharmacol. exp. Ther. **157**, 599 (1967).
233. NIEMIRO, R.: Aminoaciduria induced with maleic acid. IV. The toxic dose of maleate in vivo and in vitro. Acta biochim. pol. **7**, 95 (1960).
234. NOEDEN, G.H.: Scientific notices—chemistry, cystic oxide—communicated in a letter from Dr. Noeden to Mr. Children. Ann. Philos. **7**, 146 (1824) quoted from [363].
235. NYHAN, W.L., BORDEN, M., CHILDS, B.: Idiopathic hyperglycinemia: A new disorder of amino acid metabolism. II. The concentrations of other amino acids in the plasma and their modification by the administration of leucine. Paediatrics **27**, 539 (1961).
236. ORLOWSKI, M., MEISTER, A.: The γ-glutamyl cycle: a possible transport system for amino acids. Proc. nat. Acad. Sci. (Wash.) **67**, 1248 (1970).
237. OYANAGI KAZUHIKO, RYOICHI MIURA, TOYOSHIGE YAMANAOUCHI: Congenital lysinuria: A new inherited transport disorder of dibasic amino acids. J. Paediat. **77**, 259 (1970).
238. PASSOW, H.: Steady-state diffusion of non-electrolytes through epithelial brush borders. J. theor. Biol. **17**, 383 (1967).
239. PENG, Y., HARPER, A.E.: Amino acid balance and food intake: Effect of different dietary amino acid patterns on the plasma amino acid pattern of rats. J. Nutr. **100**, 429 (1970).
240. PERHEENTUPA, J., VISAKORPI, J.K.: Protein intolerance with deficient transport of basic amino acids. Another inborn error of metabolism. Lancet **1965 II**, 813.
241. PERRY, T., HARDWICK, D.F., LOWRY, R.B., HANSEN, S.: Hyperprolinemia in two successive generations of a north american indian family. Ann. hum. Genet. Lond. **31**, 401 (1968).
242. PFALLER, W., SILBERNAGL, S., DEETJEN, P.: Cellular localization of L-arginine reabsorption in proximal tubules of rat kidney. Pflügers Arch. **355**, R64 (1975).
243. PITTS, R.F.: A renal reabsorptive mechanism in the dog common to glycine and creatine. Amer. J. Physiol. **140**, 156 (1943).
244. PITTS, R.F.: A comparison of the renal reabsorptive processes for several amino acids. Amer. J. Physiol. **140**, 535 (1944).
245. PITTS, R.F.: Physiology of the kidney and body fluids. Chikago, Year Book, Medical Publ. Inc. (1963).
246. PITTS, R.F.: Metabolism of amino acids by the perfused rat kidney. Amer. J. Physiol. **220**, 862 (1971).
247. PROCOPIS, P.G., TURNER, B.: Iminoaciduria: A benign renal tubular defect. J. Pediat. **79**, 419 (1971).
248. PRUCANSKI, W.: Cystinuria and cystine urolithiasis in childhood. Acta paediat. (Uppsala) **55**, 97 (1966).
249. QUEHENBERGER, P., SILBERNAGL, S., DEETJEN, P.: pH-Stabilität verschiedener Puffer bei kontinuierlicher Perfusion proximaler Tubuli. Nieren- und Hochdruckkrankheiten **3**, 280 (1974).
250. REISER, S., CHRISTIANSEN, P.A.: A cross-inhibition of basic amino acid transport by neutral amino acids. Biochim. biophys. Acta (Amst.) **183**, 611 (1969).
251. REISER, S., CHRISTIANSEN, P.A.: Stimulation of basic amino acid uptake by certain neutral amino acids in isolated intestinal epithelial cells. Biochim. biophys. Acta (Amst.) **241**, 102 (1971).
252. REYNOLDS, R., REA, C., SEGAL, S.: Regulation of amino acid transport in kidney cortex of newborn rats. Science **184**, 68 (1974).
253. RICHARDS, A.N., WALKER, A.M.: Methods of collecting fluid from known regions of the renal tubules of amphibia and of perfusing the lumen of a single tubule. Amer. J. Physiol. **118**, 111 (1937).
254. ROBINSON, J.W.L., FELBER, J.P.: The absorption of dibasic amino acids by rat intestinal slices. Biochem. Z. **343**, 1 (1965).
255. ROBSON, E.B., ROSE, G.A.: The effect of intravenous lysine on the renal clearances of cystine, arginine and ornithine in normal subjects, in patients with cystinuria and fanconi syndrom and in their relatives. Clin. Sci. **16**, 75 (1957).
255a. ROSENBERG, L.E., BLAIR, A., SEGAL, S.: Transport of amino acids by slices of rat-kidney cortex. Biochim. biophys. Acta (Amst.) **54**, 479 (1961).
256. ROSENBERG, L.E., DOWNING, S.J., SEGAL, S.: Competitive inhibition of dibasic amino acid transport in rat kidney. J. biol. Chem. **237**, 2265 (1962).

257. ROSENBERG, L.E., SEGAL, S.: Maleic acid-induced inhibition of amino acid transport in rat kidney. Biochem. J. **92**, 345 (1964).
258. ROSENBERG, L.E., DURANT, J.L., HOLLAND, J.M.: Intestinal absorption and renal extraction of cystine and cysteine in cystinuria. New. Engl. J. Med. **273**, 1239 (1965).
259. ROSENBERG, L.E., SEGAL, S.: Cystinuria: Biochemical evidence for two genetically distinct diseases. J. clin. Invest. **40**, 1092 (1965).
260. ROSENBERG, L.E., DOWNING, S., DURANT, J.L., SEGAL, S.: Cystinuria: Biochemical evidence for three genetically distinct diseases. J. clin. Invest. **45**, 365 (1966).
261. ROSENBERG, L.E., CRAWHALL, J.C., SEGAL, S.: Intestinal transport of cystine and cysteine in man: Evidence for separate mechanisms. J. clin. Invest. **46**, 30 (1967).
262. ROSENBERG, L.E., DURANT, J.L., ELSAS, L.J.: Familial iminoglycinuria. An inborn error of renal tubular transport. New. Engl. J. Med. **278**, 1407 (1968).
263. ROSENHAGEN, M., SEGAL, S.: Stereospecificity of amino acid uptake by rat and human kidney cortex slices. Amer. J. Physiol. **227**, 843 (1974).
264. RUMRICH, G., ULLRICH, K.J.: The minimum requirements for the maintenance of sodium chloride reabsorption in the proximal convolution of the mammalian kidney. J. Physiol. (Lond.) **197**, 69P (1968).
265. RUSSO, H.F., WRIGHT, H.R., SKEGGS, H.R., TILLSON, E.K., BEYER, K.H.: Renal clearance of essential amino acids: Threonine and phenylalanine. Proc. Soc. exp. Biol. (N.Y.) **65**, 215 (1947).
266. RUSZKOWSKI, M., BAERTL, J.M., GABUZDA, G.J.: Cystinuria-like derangement of amino acid excretion in patients with hepatic cirrhosis given arginine intravenously. J. clin. Invest. **38**, 1038 (1959).
267. RUSZKOWSKI, M., ARASIMOWICZ, C., KNAPOWSKI, J., STEFFEN, J., WEISS, K.: Renal reabsorption of amino acids. Amer. J. Physiol. **203**, 891 (1962).
268. SANO, KINGO: Über die Löslichkeit der Aminosäuren bei variierter Wasserstoffzahl. Biochem. Z. **168**, 14 (1926).
269. SCHAFER, I.A., SCRIVER, C.R., EFRON, M.L.: Familial hyperprolinemia, cerebral dysfunction and renal anomalies occurring in a family with hereditary nephropathy and deafnes. New. Engl. J. Med. **267**, 51 (1962).
270. SCHNERMANN, J., HORSTER, M., LEVINE, D.Z.: The influence of sampling technique on the micropuncture determination of GFR and of single rat proximal tubules. Pflügers Arch. **309**, 48 (1969).
271. SCHNERMANN, J., DAVIS, J.M., WUNDERLICH, P., LEVINE, D.Z., HORSTER, M.: Technical problems in the micropuncture determination of nephron filtration rate and their functional implications. Pflügers Arch. **329**, 307 (1971).
272. SCHREIER, K., MÜLLER, W.: Idiopathische Hyperglycinämie (Glycinose). Dtsch. med. Wschr. **89**, 1739 (1964).
273. SCHWARTZMAN, L., BLAIR, A., SEGAL, S.: Exchange diffusion of dibasic amino acids in the rat kidney cortex slices. Biochim. biophys. Acta (Amst.) **135**, 120 (1967).
274. SCRIVER, C.R., SCHAFER, I.A., EFRON, M.L.: New renal tubular amino acid transport system and new hereditary disorder of amino acid metabolism. Nature (Lond.) **192**, 672 (1961).
275. SCRIVER, C.R., EFRON, M.L., SCHAFER, I.A.: Renal tubular transport of proline, hydroxyproline and glycine in health and familial hyperprolinemia. J. clin. Invest. **43**, 374 (1964).
276. SCRIVER, C.R.: Hartnup disease. A genetic modification of intestinal and renal transport of certain neutral alpha-amino acids. New Engl. J. Med. **273**, 530 (1965).
277. SCRIVER, C.R., GOLDMAN, H.: Renal tubular transport of proline, hydroxyproline and glycine. II. Hydroxy-L-proline as substrate and inhibitor in vivo. J. clin. Invest. **45**, 1357 (1966).
278. SCRIVER, C.R., WILSON, O.H.: Amino acid transport: Evidence for genetic control of two types in human kidney. Science **155**, 1428 (1967).
279. SCRIVER, C.R.: Renal tubular transport of proline, hydroxyproline and glycine. III. Genetic basis for more than one mode of transport in human kidney. J. clin. Invest. **47**, 823 (1968).
280. SCRIVER, C.R., MOHYUDDIN, F.: Amino acid transport in kidney. Heterogeneity of alpha-aminoisobutyric uptake. J. biol. Chem. **243**, 3207 (1968).
281. SEGAL, S., BLAIR, A., ROSENBERG, L.E.: The effect of phlorizin on amino acid transport in rat-kidney-cortex slices. Biochim. biophys. Acta (Amst.) **71**, 676 (1963).

282. SEGAL, S., CRAWHALL, J.C.: Transport of cysteine by human kidney cortex in vitro. Biochem. Med. **1**, 141 (1967).
283. SEGAL, S., SMITH, I.: Delineation of separate transport systems in rat kidney cortex for L-lysine and L-cystine by developmental patterns. Biochem. biophys. Res. Commun. **35**, 771 (1969).
284. SEGAL, S., SMITH, I.: Delineation of cystine and cysteine transport systems in rat kidney cortex by developmental patterns. Proc. nat. Acad. Sci. (Wash.) **63**, 926 (1969).
285. SEGAL, S., THIER, S.O.: Renal handling of amino acids. In: (ORLOFF, J., BERLINER, R.W., eds.) Handbook of Physiology, vol. 8: Renal Physiology. Am. Physiol. Soc. Washington D.C. (1973).
286. SELKOE, D.J.: Familial hyperprolinemia and mental retardation. A second metabolic type. Neurology (Minneap.) **19**, 494 (1969).
287. SELS, A.: Analogues of amino acids. In: Fundamentals of biochemical pharmacology. (BACQ, Z.M., ed.), p. 505, Oxford-New York-Toronto-Sydney-Braunschweig: Pergamon Press 1971.
288. SERENI, F., MCNAMARA, H., SHIBUYA, M., KRETCHMER, N., BARNETT, H.L.: Concentration in plasma and rate of urinary excretion of amino acids in premature infants. Pediatrics **15**, 575 (1955).
289. SHALHOUB, R., WEBBER, W., GLABMAN, S., CANESSA-FISCHER, M., KLEIN, J., DEHAAS, J., PITTS, R.F.: Extraction of amino acids from and their addition to renal blood plasma. Amer. J. Physiol. **204**, 181 (1963).
290. SHAW, K.N.F., BODER, E., GUTENSTEIN, M., JACOBS, E.E.: Histidinemia. J. Pediat. **63**, 720 (1963).
291. SHIK, V.E., BIXBY, F.M., ALPERS, D.H., BARTSOCAS, C.S., THIER, S.O.: Studies of intestinal transport defect in Hartnup disease. Gastroenterology **61**, 445 (1971).
292. SHIPP, J., HANENSON, I., WINDHAGER, E.E., SCHATZMANN, H., WHITTENBURY, G., YOSHIMURA, H., SOLOMOM, A.K.: Single proximal tubules of Necturus kidney. Methods for micropuncture and microperfusion. Amer. J. Physiol. **195**, 563 (1958).
293. SILBERNAGL, S., DEETJEN, P.: Mikroperfusionsuntersuchungen zur Resorption von Gycin im proximalen Tubulus. Pflügers Arch. **312**, R82 (1969).
294. SILBERNAGL, S.: Mikroperfusionsuntersuchungen zum tubulären Aminosäuren-Transport. In: Fortschritte der Nephrologie. (BOHLE, A., SCHUBERT, G.E., Eds.), p. 157. Stuttgart-New York: Schattauer 1971.
295. SILBERNAGL, S., DEETJEN, P.: Glycine reabsorption in rat proximal tubules. Microperfusion studies. Pflügers Arch. **323**, 342 (1971).
296. SILBERNAGL, S., DEETJEN, P.: Amino acid transport in rat proximal tubule. Proceedings of the International Union of Physiological Sciences, **9**, p. 514: XXV. International Congress, München 1971.
297. SILBERNAGL, S.: Mikropunktionsuntersuchungen zum tubulären Transport der "basischen" Aminosäuren an der Ratte. In: HEINTZ, R., HOLZHÜTER, H.: Renale Elimination von Pharmaka. Immunologie und Klinik der Nierentransplantation. Zur Pathophysiologie des Dialyse-Patienten. Regeneration der Niere. VIII. Symposion der Gesellschaft für Nephrologie, p. 649, Aachen (1972).
298. SILBERNAGL, S.: Specifity of the L-arginine transport in rat proximal tubule. Pflügers Arch. **332**, (Suppl.), R30 (1972).
299. SILBERNAGL, S.: Influence of L-arginine on the L-cystine transport in the rat proximal tubule. Pflügers Arch. **335**, (Suppl.), R40 (1972).
300. SILBERNAGL, S., DEETJEN, P.: L-arginine transport in rat proximal tubules. Microperfusion studies on reabsorption kinetics. Pflügers Arch. **336**, 79 (1972).
301. SILBERNAGL, S., DEETJEN, P.: The tubular reabsorption of L-cystine and L-cysteine. A common transport system with L-arginine or not? Pflügers Arch. **337**, 277 (1972).
302. SILBERNAGL, S., DEETJEN, P.: Handling of L-arginine and of its derivatives in rat proximal tubule. Microperfusion study. Abstracts of V. International Congress of Nephrology, Mexico City, p. 130 (1972).
303. SILBERNAGL, S., DEETJEN, P.: Molecular specificity of the L-arginine reabsorption mechanism. Microperfusion studies in the proximal tubule of rat kidney. Pflügers Arch. **340**, 325 (1973).
304. SILBERNAGL, S.: Some problems of L-cystine transport in the kidney tubule of rat. Pflügers Arch. **343**, R46 (1973).
305. SILBERNAGL, S.: Physiologie und Pathophysiologie der Aminosäurenresorption in der Niere. Habilitationsschrift, Innsbruck (1973).

306. SILBERNAGL, S.: Die tubuläre Reabsorption einiger Transport-Hemmstoffe und deren Wirkung auf den Aminosäuren-Transfer in der Rattenniere. Nieren- und Hochdruckkrankheiten, **3**, 18 (1974).

307. SILBERNAGL, S.: The effect of penicillamine and L-homoserine on L-cystine transport in rat kidney. Pflügers Arch. **347**, R69 (1974).

308. SILBERNAGL, S.: Handling of cycloleucine (1-amino-cyclopentane carboxylic acid) in the proximal tubule of rat kidney. INSERM, Colloque Europeen: Physiologie du néphron: Mécanisme et régulation, 193 (1974).

309. SILBERNAGL, S.: Aminosäuren-Transport in der Niere—Ergebnisse der Mikroperfusion. BIUZ **4**, (6), 161 (1974).

310. SILBERNAGL, S.: Cycloleucine (1-amino-cyclopentane carboxylic acid): tubular reabsorption and inhibitory effect on amino acid transport in the rat kidney (microperfusion experiments). Pflügers Arch. **353**, 241–253 (1975).

311. SILBERNAGL, S.: Renal handling of amino acids. Clinical Nephrology, in press (1976).

312. SILBERNAGL, S., QUEHENBERGER, P., MAREN, T.H.: Proton permeability of the proximal tubule of the rat kidney. Pflügers Arch. **355**, R54 (1975).

312a. SILBERNAGL, S.: Renal tubular reabsorption of amino acids: Specificity of the different transport systems studied by continuous microperfusion. In: Amino acid transport (ed. SILBERNAGL, S.) and uric acid (ed. LANG, F., GREGER, R.). Stuttgart: Thieme 1975 (in press).

312b. SILBERNAGL, S.: Renal handling of L-Methionine and other neutral amino acids studied by continuous microperfusion. Pflügers Arch. **359**, R 120 (1975).

313. SILBERNAGL, S.: unpublished results.

314. SILK, D.B.A., PERRETT, D., STEPHENS, A.D., CLARK, M.L., SCOWEN, E.F.: Intestinal absorption of cystine and cysteine in normal human subjects and patients with cystinuria. Clin. Sci. **47**, 393–397 (1974).

315. SILVERMAN, M., AGANON, M.A., CHINARD, F.P.: Specificity of monosaccharide transport in dog kidney. Amer. J. Physiol. **218**, 743 (1970).

316. SIMILÄ, S., VISAKORPI, J.K.: Hyperprolinemia without renal desease. Acta paediat. (Uppsala) **177**, (Suppl.), 122 (1967).

317. SOBER, H.A. (ed.): Handbook of biochemistry, selected data for molecular biology. The Chemical Rubber Co. 2nd ed. Cleveland/Ohio (1970).

318. SOLOMON, S.: Method for investigating net H_2O fluxes across individual proximal tubules. Proc. Soc. exp. Biol. (N.Y.) **101**, 221 (1959).

319. SONNENBERG, H., DEETJEN, P.: Methode zur Durchströmung einzelner Nephronabschnitte. Pflügers Arch. ges. Physiol. **278**, 669 (1964).

320. SPENCER, A.G., FRANGLEN, G.T.: Gross aminoaciduria following a lysol burn. Lancet **1952** **I**, 190.

321. SPITZER, A., WINDHAGER, E.E.: Effect of peritubular oncotic pressure changes on proximal tubular fluid reabsorption. Amer. J. Physiol. **218**, 1188 (1970).

322. SPITZER, A., WINDHAGER, E.E.: Continuous in vivo perfusion of the postglomerular capillary network in superficial rat kidney cortex. Yale J. Biol. Med. **45**, 307 (1972).

323. STALDER, G., VETTERLI-BUCHNER, H., BERGER, H.: Untersuchungen über den renalen Rückresorptionsmechanismus für Aminosäuren. Klin. Wschr. **38**, 278 (1960).

324. STEIN, W.H.: Excretion of amino acids in cystinuria. Proc. Soc. exp. Biol. (N.Y.) **78**, 705 (1951).

325. STEIN, W.S., MOORE, S.: The free amino acids of human blood plasma. J. biol. Chem. **211**, 915 (1954).

326. STERN, J.R., EGGLESTON, L.V., HEMS, R., KREBS, H.A.: Accumulation of glutamino acid in isolated brain tissue. Biochem. J. **44**, 410 (1949).

327. SUGITA, M., SUGITA, K.O., FURKAWA, T., ABE, H.: Studies on the transport mechanism of amino acids in the renal tubules. I. Studies on the mechanism of aminoaciduria from the analytical standpoint of titration curve. Jap. Circulat. J. **31**, 405 (1967).

328. TADA KEIYA, TOSHIO MORIKAWA, TOSHIYUKI ANDO, TOSHIO YOSHIDA, AKIBUMI MINAGAWA.: Prolinuria: A new renal tubular defect in transport of proline and glycine. Tohoku J. exp. Med. **87**, 133 (1965).

329. TANCREDI, F., GUAZZI, G., AURICCHIO, S.: Renal iminoglycinuria without intestinal malabsorption of glycine and imino acids. J. Pediat. **76**, 386 (1970).

330. THIER, S., FOX, M., SEGAL, S.: Cystinuria: In vitro demonstration of an intestinal transport defect. Science **143**, 482 (1964).
331. THIER, S., FOX, M., ROSENBERG, L.E., SEGAL, S.: Hexose inhibition of amino acid uptake in rat-kidney-cortex slice. Biochim. biophys. Acta (Amst.) **93**, 106 (1964).
332. THIER, S., SEGAL, S., FOX, M., BLAIR, A., ROSENBERG, L.E.: Cystinuria: Defective intestinal transport of dibasic amino acids and cystine. J. clin. Invest. **44**, 442 (1965).
333. THIER, S., BLAIR, A., FOX, M., SEGAL, S.: The effect of extracellular sodium concentration on the kinetics of amino-isobutyric acid transport in the rat kidney cortex slice. Biochim. biophys. Acta (Amst.) **135**, 300 (1967).
334. THIER, S.: Inborn errors of organic solute transport: Genetic control of amino acid transport in gut and kidney. Birth defects **6**, 20 (1970).
335. THIER, S., SEGAL, S.: Cystinuria. In: The metabolic basis of inherited disease. 3rd ed. by STANBURY, J.B., WYNGARDEN, J.B., FREDERICKSON, D.S. Mc Graw-Hill 1504 (1972).
336. TREACHER, R.J.: Amino acid excretion in canine cystine-stone disease. Vet. Rec. **74**, 503 (1962).
337. TREACHER, R.J.: The aetiology of canine cystinuria. Biochem. J. **90**, 494 (1964).
338. TSAN MIN-FU, JONES, T.C., THORTON, G.W., LEVY, H.L., GILMORE, G., WISON, T.H.: Canine cystinuria: Its urinary amino acid pattern and genetic analysis. Amer. J. Vet. Res. **33**, 2455 (1972).
339. TSAN MIN-FU, JONES, T.C., WILSON, T.H.: Canine cystinuria: Intestinal and renal amino acid transport. Amer. J. vet. Res. **33**, 2463 (1972).
340. ULLRICH, K.J., FRÖMTER, E., BAUMANN, K.: Micropuncture and microanalysis in kidney physiology. In: Laboratory techniques in membrane biophysics. (PASSOW, H., STÄMPFLI, R., eds.), p. 106. Berlin-Heidelberg-New York: Springer 1969.
341. ULLRICH, K.J., RUMRICH, G., KLÖSS, S.: Sodium dependence of the transtubular transport of glucose and amino acid in the proximal tubule of the rat kidney. Pflügers Arch. **339**, (Suppl.) R47 (1973).
342. ULLRICH, K.J., RUMRICH, G., KLÖSS, S.: Sodium dependence of the amino acid transport in the proximal convolution of the rat kidney. Pflügers Arch. **351**, 49–60 (1974).
343. ULLRICH, K.J., FRÖMTER, E., EVERS, J., KINNE, R.: Sodium dependence of the amino acid transport in the proximal convolution of the rat kidney. In: Amino acid transport (ed. SILBERNAGL, S.) and uric acid (ed. LANG, F., GREGER, R.). Stuttgart: Thieme 1975 (in press).
344. DEVRIES, A., KOCHWA, S., LANZENBIK, J., FRANK, M., DJALDETTI, M.: Glycinuria, a hereditary disorder associated with nephrolithiasis. Amer. J. Med. **23**, 408 (1957).
345. WAISMAN, H.A.: Variation in clinical and laboratory findings in histidinemia. Amer. J. Dis. Child **113**, 93 (1967).
346. WEBBER, W.A., BROWN, J.L., PITTS, R.F.: Interactions of amino acids in renal tubular transport. Amer. J. Physiol. **200**, 380 (1961).
347. WEBBER, W.A.: Interactions of neutral and acidic amino acids in renal tubular transport. Amer. J. Physiol. **202**, 577 (1962).
348. WEBBER, W.A.: Characteristics of acidic amino acid transport in mammalian kidney. Canad. J. Biochem. **41**, 131 (1963).
349. WEBBER, W.A., CAMPELL, J.L.: Effects of amino acids on renal glucose reabsorption in the dog. Canad. J. Physiol. Pharmacol. **43**, 915 (1965).
350. WEBBER, W.A.: Some effects of phlorizin and phloretin on renal amino acid reabsorption in the dog. Canad. J. Physiol. Pharmacol. **43**, 79 (1965).
351. WEBBER, W.A.: Renal tubular reabsorption of α-aminoisobutyric acid. Canad. J. Physiol. Pharmacol. **44**, 507 (1966).
352. WEDEEN, R.P., WEINER, B.: Distribution of PAH in rat kidney slices II. Depth of uptake. Kidney Internatl. **3**, 214 (1973).
353. WEINMANN, E.J., HARDY, R.J., KASHGARIAN, M., HAYSLETT, J.P.: Examination of the Gertz technique as applied to the proximal tubule of the rat kidney. Yale J. Biol. Med. **45**, 289 (1972).
354. WEISE, M., EISENBACH, G.M., STOLTE, H.: Mikropunktionsuntersuchungen über die Resorption freier Aminosäuren im Einzelnephron der Säugetierniere. In: Fortschritte der Nephrologie, Niere und Stoffwechsel, Juxta-glomerulärer Apparat. VII. Symposion der Gesellschaft für Nephrologie, Tübingen (1970). (BOHLE, A., SCHUBERT, G.F., eds.), p. 161. Stuttgart-New York: Schattauer 1971.

355. WEISE, M., EISENBACH, G.M., STOLTE, H.: Mikropunktionsuntersuchungen über die renale Aminosäurenresorption. In: Renale Elimination von Pharmaka, Immunologie und Klinik der Nierentransplantation. Zur Pathophysiologie des Dialyse-Patienten. Regeneration der Niere. VIII. Symposion, Gesellschaft für Nephrologie, Aachen (1971). (HEINTZ, R., HOLZHÜTER, H. eds.), p. 641. Aachen 1972.
356. WEYERS, H., BICKEL, H.: Photodermatose mit Aminoacidurie, Indolaceturie und cerebralen Manifestationen (Hartnup-Syndrom). Klin. Wschr. **36**, 839 (1958).
357. WHELAN, D.T., SCRIVER, C.R.: Hyperdibasicaminoaciduria: An inherited disorder of amino acid transport. Pediat. Res. **2**, 525 (1968).
358. WHELAN, D.T., SCRIVER, C.R.: Cystathioninuria and renal iminoglycinuria in a pedigree. A perspective on counseling. New Engl. J. Med. **278**, 924 (1968).
359. WILLIAMS, W.M., HUANG, K.C.: In vitro and in vivo renal tubular transport of tryptophan derivatives. Amer. J. Physiol. **219**, 1468 (1970).
360. WILSON, V.K., THOMSON, M.L., DENT, C.E.: Aminoaciduria in lead poisoning. A case in childhood. Lancet **1953 II**, 66.
361. WILSON, O.H., SCRIVER, C.R.: Specificity of transport of neutral and basic amino acids in rat kidney. Amer. J. Physiol. **213**, 185 (1967).
362. WINDHAGER, E.E.: Micropuncture techniques and nephron function. London: Butterworth 1968.
363. WIRTSCHAFTER, Z.T.: The role of metals in carbohydrate metabolism. J. Lab. clin. Med. **26**, 1093 (1941).
364. WOLLASTON, W.H.: On cystic oxide: A new species of urinary calculus. Trans. roy. Soc. Edinb. **100**, 223 (1810).
365. WOODY, N.C., SNYDER, C.H., HARRIS, J.A.: Histidinemia. Amer. J. Dis. Child **110**, 606 (1965).
366. WOODY, N.C., SNYDER, C.H., HARRIS, J.A.: Hyperprolinemia: Clinical and biochemical family study. Paediatrics **44**, 554 (1969).
367. WORTHEN, H.G.: Renal toxicity of maleic acid in the rat. Lab. Invest. **12**, 791 (1963).
368. WRIGHT, L.D., RUSSO, H.F., SKEGGS, H.R., PATCH, E.A., BEYER, K.H.: The renal clearance of essential amino acids: Arginine, lysine and methionine. Amer. J. Physiol. **149**, 130 (1947).
369. YEH HUI LAN, FRANKL, W., DUNN, M.S., HUGHES, B., GYÖRY, P.: The urinary excretion of amino acids by a cystinuria subject. Amer. J. med. Sci. **214**, 507 (1947).
370. YOUNG, J.A., EDWARDS, K.D.G.: Studies on the absorption metabolism and excretion of methyldopa and other catechols and their influence on amino acids transport in rats. J. Pharmacol. exp. Ther. **145**, 102 (1964).
371. YOUNG, J.A., EDWARDS, K.D.G.: Clearance and stop-flow studies on histidine and methyldopa transport by rat kidney. Amer. J. Physiol. **210**, 667 (1966).
372. YOUNG, J.A., FREEDMAN, B.S.: Renal tubular transport of amino acids. Clin. Chem. **17**, 245 (1971).

Author Index

Page numbers in *italics* refer to the bibliography. Numbers shown in square brackets are the numbers of the references in the bibliography

Subject Index